an engineer of sorts

The Author receiving the double Queen's Award to Industry from the Lord Lieutenant of Hampshire 1966

Peter Du Cane

C.B.E., C.Eng. F.R.I.N.A., F.I.Mech.E.
M.I.Mar.E., A.F.R.Ae.S., Royal Navy

an engineer of sorts

Erroll Bruce
Richard Creagh-Osborne
Peter Johnson

Hon. President:
Adlard Coles

Nautical Publishing Company

Nautical House
Lymington, Hampshire

in association with George G. Harrap & Co. Ltd.,
London, Toronto, Sydney, Wellington

First Published in Great Britain by
NAUTICAL PUBLISHING COMPANY
Nautical House, Lymington, Hampshire

ISBN 0 245 50749 3

Photoset and printed in Great Britain by
BAS Printers Limited, Wallop, Hampshire

Dedicated to Douglas Phillips-Birt who has often inspired me towards the more interesting side of naval architecture and whom I regard as a loyal friend, whose work I admire.
Also to my friends at Vosper of all categories, without whom what could be called a successful modern shipyard could not have been made.

Acknowledgements

To all the many who have helped me through my career in and out of Vosper. Also to my family whose patience and suggestions have been of great help.

In particular I owe a debt of gratitude to Miss Jean Carpenter my secretary who has typed draft and fair copy of the script and whose help throughout has been invaluable.

Contents

Illustrations

Foreword

While there are plenty of books about or by people who use ships, there are few concerning the builders and designers of ships. This is one reason to welcome Peter Du Cane's autobiography. Another is that his work as a designer and builder has been chiefly concerned with the most fascinating type of marine craft in the modern age—the high speed relatively small vessels that became possible when the petrol engine was first put afloat, and which since then have been in the van of developments in marine engineering, as well as the fastest craft ever produced.

The small band of people who were involved with such craft began creating a stir in the years prior to 1914, and the "engineer of sorts" joined them after the war. Apart from this chief interest, which brought him international fame as a designer and as Managing Director of Vosper (now Vosper Thornycroft) the immensely varied and active life of the author has freed him from the deadening effect of becoming over-specialised. Naval officer, yachtsman, engineer, airman, enthusiast for fast cars, hunting man, naval architect, and one who appreciates the amenities of civilised life—these widely ranging activities ensure both entertainment and instruction in the story he has to tell.

Some of these activities gave him unique qualifications for the exciting branch of shipbuilding which became his main occupation. It is said, not always without reason, that the makers of ships are liable to be out of touch with the users of ships. The author was a practical handler of boats from an early age, and has always remained one. He also gained the high technical qualifications of a naval engineer. His flying experience might have had marginal value only for the average naval architect, but it is unquestioned nowadays that flying experience and "know-how" may be of more value to the designer of high speed planing craft than much of the corpus of basic naval architecture.

Peter Du Cane has combined practical affairs with a continuous, lively interest in theory. Thus while running a shipyard and being involved in the personal contacts and negotiations entailed in the management and business spheres (and some of the stories he has to tell here are amusing) he devoted much time to research on hulls and propellers, and for the latter established excellent facilities at Vosper. His interest in hydrodynamics is revealed in a most readable way in these pages, and more technically appears in the various papers he has read, and continues to read, to learned societies in the U.K. and abroad; also in his previous book, shortly to reappear in its 4th edition. The general reader will find particular interest in his account of his connection with the design and construction of the Unrestricted Class World's Speed Record craft, and later with offshore powerboats. His name is particularly associated with Sir Malcolm Campbell's *Bluebird*, which held the World's Water Speed Record between 1939 and 1950, and with *Tramontana*, winner of the second Daily Express International Offshore Power Boat race in 1962. Long experience with such temperamental and often unpredictable craft inevitably brought disappointments too: the tragedy to John Cobb's *Crusader* at a speed that would have brought her the World's Record in 1952; the loss, evidently through striking a log, of *Flying Fish* when she was leading the International Offshore race of 1966. Such record breaking and racing activities were an aside from the continuous work of developing fast naval craft, which reached its climax, as told in these pages, during 1939–45.

It might seem pompous, in a book so happily informal, to say it is a privilege to contribute a preface. It has certainly been of pleasure and instruction to read the story it tells.

Douglas Phillips-Birt

1

Early Days and the Navy

Possibly, the fact that I was sent for the holidays from the age of about five to Newbrough in Northumberland without my parents had a bit to do with my becoming an engineer. My cousins, who lived about a mile away, owned a number of coal mines and quarries in the near vicinity, and I had a superfluity of energy and inquisitiveness even at this age.

A few of us, who later become class conscious, do not have much of this in our make up at an early age, it was not long before a bicycle helped me to get down to the local pit at Fourstones on Tyne, where apart from riding on the tubs up and down I did many stints of an hour or so hewing at the coal face, much to the amusement of the real miners.

Of course, if my cousins had known of this it would have been stopped at once, but rather miraculously they did not—at least for some years—although my grandparents, with whom I was staying, were most annoyed at the coating of coal dust with which I turned up to breakfast most days. Things came to a climax, however, when assisted by my cousin, who was about my age, we managed to get some trucks in the quarry near-by to the edge of a precipice and jack them over. This was a bit too much, and being seen by someone who reported it my activities were curbed for at least a month.

There were so many wonderful things for a boy to do there, including driving—quite without authority—one of the new motor cars of about 1914, which was chauffeured for my grandparents by a girl, owing to the call up. I jumped into the driving seat when she was temporarily out of it, and she, losing her head, managed to distract me enough to allow the car to mow down a stone pelican and a length of ornamental rails which adorned the front entrance of our important residence. Quite unforgivable, of course, on my part; but I had in fact driven an enormous Napier about two years before in the park of one of the Rothschild's, with the permission of an extremely sporting and cooperative chauffeur. I think I was then 10 years old, certainly not much more.

Anyway, it seems that mechanical devices of all sorts had a fascination for me, even to the extent of my taking a great interest in lifting the lids of drainage systems and tracing the drain piping to its source—usually a cess-pit—in any house in which I was lucky enough to be invited. The water rams in the woods at Newbrough were a source of great satisfaction.

My cousins had one of the first electricity systems to light the house, which involved a heavy slow running and belt driven oil engine and dynamo in which I managed to get caught up and flung against the railings one day. This caused much of a stir because my grandfather, while a man of much rectitude, had little humour and took quite seriously a remark I made while I was being carried into the house with a pretty badly broken arm, that this would prevent me returning to prep school for some time. I even got caned when the arm was better, but my principal recollection of this episode was the suffocating feeling of having a mask and chloroform held over my face while the doctor set the arm. This was 1909 before the pentothal technique.

My first experience of the Royal Navy came about 1912 in Bantry Bay where as a prep schoolboy I used to stay with my mother at Bantry House. At the time a house party was being entertained in a way, impossible nowadays, by the hospitable Mrs. Leigh White and her husband. One evening the officers from HMS *Lord Nelson*, laying in the harbour, invited me to take a run in their steam picket boat to visit the ship. Even at this stage I found the little engine room and forced draught boiler room fascinating. Of course, the kind hearted stoker petty officer allowed me to operate the reciprocating steam engine, which kept me busy for a matter of days on end, and doubtless relieved the Bantry House party of the tiresomeness of at least one small boy. I do recollect, however, that in the meantime, or before discovering this fascinating picket boat, I had managed to rig up an aerial railway between two trees in the grounds at Bantry House; an overhead rope railway which was much to the delight of the two young ladies of about my age who were the daughters of the house.

At any rate, so far as I can remember, this visit of the *Lord Nelson* to Bantry confirmed my decision to join the Royal Navy as soon as I was old enough and if they would have me. There was also a little matter of the interview and the exams to be passed. Strange to say I passed into Osborne in 1914 in fourth place out of more than 100 cadets. This did not, however, prevent my term officer during my last term at Dartmouth telling me in the kindest way that it had recently been established that I had received more corporal punishment than any cadet up to date since the college had opened!

There was plenty of sailing at Dartmouth in the Black Cutters as they were called in those days and at that time I opted for the river rather than the cricket and rugger fields.

We also had the chance to run the machinery and take the instructional steam picket boat up and down the Medina as cadets at Osborne, aged 14. I would collect four or five kindred spirits and during the week-end or silent hours get her steam up and pay a visit to the *Tipperary*, a destroyer building at J. S. White's yard—I think to their own design. Another never failing attraction was the topsail schooner *Valhalla* belonging to Sir Joseph Laycock, but laid up for the war period. We used to lay out along the yards as good practice but, of course, were inevitably seen by the authorities from Kingston the engineering workshops to which the steam boat belonged. Hardly surprisingly I was beaten for the double offence of raising steam in the picket boat without permission and boarding the *Valhalla*, which was out of bounds to all cadets.

When at sea in HMS *St. Vincent* during 1917 I always went down to one of the boiler rooms and shovelled coal on the furnaces during the dog watches, which was really a most satisfactory task for a little while. The skill required was quite considerable; also the strength and endurance. But I was young—16–17—and the real stokers were as kind and amused as you could expect. Probably most of my instruction came from watching them. They were kindness itself.

For some reason—I cannot imagine why—I did a lot of stoking in a passenger liner (coal) on my way out from Liverpool to Malta a few years later to join HMS *Malaya*. I do not think I would do it now, but in those days one was keen to keep fit, which has always seemed a necessity to me.

Joining HMS *Revenge* as a midshipman in 1919 I found myself as senior midshipman having served during and just after the war in HMS *St. Vincent* and then HMS *Whitby*, a destroyer of the "W" class. Being senior midshipman enabled me to allocate myself to what I considered to be the plum job, that is to say running in charge of the steam picket boats, of which there were two.

In those days the midshipman always handled the boat himself, which does not seem to happen to-day. Doubtless a few more rubbing strakes were split and gangways broken, but it was a wonderful experience for a young officer who, if he stays long enough to become proficient, will never lose the knack of handling boats of all sorts. The dipping lug cutters and Montague whalers gave us our first experience of sail—that is to say after the black cutters of the Dartmouth days.

Strangely enough, it was during peacetime in HMS *Revenge* that I received most of my picket boat experiences of any real interest. Though a battleship spending much time in Scapa Flow, *St. Vincent* tended to be at sea a good deal, especially towards the end of the war after the German High Seas Fleet had sunk the *Mary Rose* and *Strongbow* while escorting a convoy of merchant ships off the Norwegian coast.

In the summer of 1919 we embarked upon a cruise of the eastern coast of England calling, amongst other ports, at Yarmouth where I had the misfortune to collide with the Admiral's barge while endeavouring to come alongside the pier. This sounds like a piece of very poor seamanship, but there was an element of excuse for me, and for those not immediately involved, an element of humour.

The Flag Lieutenant, while embarking on board my picket boat at the port gangway, told me to hurry up inshore so as to get there before the Admiral's barge, which I was given to understand had only some of the Admiral's lady friends aboard and not the Admiral himself.

As we broke out clear of the stem of the ship sure enough there was the barge which had left the starboard gangway at about the same time. Unable to see the affirmative disc on the funnel of the barge, which was beam on to us, I told my engine room crew to open up to full steam ahead and we raced the barge neck and neck for the pier. As I was on the inside berth I was in a position to get alongside the pier first.

Suddenly and to my horror the coxswain of the barge came in right alongside me and in fact collided fairly violently. At this awful moment the Admiral himself, Sir Sydney Fremantle, appeared in the stern sheets and made a jump for our deck, informing me at the same time that I was to return to the ship and report to the Captain, who would place me under arrest. My punishment was watch on watch (i.e. four hours on duty and four hours off duty) for one week.

Later in the same commission we went out to Constantinople where there was trouble with Attaturk, who was annexing the whole of Asia Minor, including Armenia. This was quite against the armistic terms which envisaged the defeated Turks being occupied by an Allied commission.

One day while lying at Ismid we were told that we were to steam through the night alongside the *Revenge* with the idea of making a surprise dawn attack the next day on the Nationalists, who had occupied Mudania, a port on the western side of the gulf. Our job in the picket boat was to escort the landing parties and then patrol up and down the beach firing on any concentration of hostile troops. This we did on more than one occasion while the landing parties were being put ashore under a heavy barrage from our 15 inch guns. All was over and the town cleared of enemy pockets of resistance by early afternoon when the Admiral, again Sir Sydney, came ashore in our picket boat to proclaim the victory of our forces.

We waited for him for an hour or more and on return to the ship most of my crew went below for a rest, while I was left at the wheel. Feeling a bit sleepy I stepped over the coaming from the steering position and managed with a lurch of the boat to slip over the side into the water with the picket boat going on at full speed. Luckily the stern sheet man saw my plight and rushed with a life buoy which he jammed on to my shoulders almost causing me to let go my precarious hold on the gunwale. However, I eventually persuaded him to pull me inboard none the worse except for being wet through. On this occasion the Admiral did not notice a thing when he returned aboard, which was lucky.

The Royal Navy did me well—probably better than I did them. It was such a relief to get away from school, and even Osborne and Dartmouth, to join a battleship during the war in 1917 at the age of $16\frac{1}{2}$ that really one never looked back until later in peacetime when I, for one, found it dull; though I think if I had stuck it out for a bit longer it would have become more interesting and rewarding.

Fig 1 The author as a Midshipman

Fig 2 Brescia Bugatti Racing Car

The highlights in my Service career were, I suppose, about 3 years running a steam picket boat, which is really wonderful experience, especially as these boats had only one propeller, and called for quite a lot of skill in handling alongside; taking charge of the gunnery transmitting station while still a midshipman in wartime; and qualifying in engineering at Keyham and Greenwich to about M.Sc. standard if one had been asked to write a thesis—and a most useful course it has turned out to be in after life!

Actually I did not qualify at the final exams at Greenwich. A pal of mine died in the house of a kind lady friend during a week while the exams were on and I did not really feel I could leave her alone to make all the arrangements; in consequence I missed three papers and was given no compensation therefore—possibly justifiably—but I thought a bit hard. Anyway they did not seem to hold it against me because I was appointed to what was then called the Glass House, where all the research and back room stuff was done in the Admiralty engineering department in those days.

I am sorry to say my treatment over the exams may not have been entirely due to academic and disciplinary reasons. I was not popular with the officer in charge because I was driving to London almost every evening in a racing Bugatti and arriving covered in mud or oil or both all over a white stiff shirt and white tie. Also what I believed are now called the "bums" called at the porter's lodge most days as I was a bit short of cash, and the police were pretty frequently serving summonses on me for motoring offences—luckily at that time not particularly serious ones. But there were a lot of them.

The engineering course and Greenwich generally was good fun and has always been most valuable to me in naval and civilian life, mostly in design work as applied to both engineering and naval architecture. Contrary to what a lot may still think these are both essential disciplines as a joint piece of engineering—certainly in the faster, smaller craft where present problems lie closer to aeronautical engineering than the run of the mill naval architecture.

Fig 3 H.M.S. *Emerald*

The Royal Navy was good grounding for someone who eventually took up my sort of career, because at an early age one had responsibility in charge of small craft, both steam, motor and sail, with no one to get you there and back but yourself, and in all weathers. This was invaluable and one got to know the score—at least if you liked the sort of thing. Strangely enough it had a valuable side effect because if you cannot get on with the ship's company you have not really a chance of "throwing the discipline act on them" and getting away with it. This is an intangible quality which some have and others never obtain, but it makes just all the difference. Mostly, I suppose, the men look up to and get on with someone who is obviously competent, possesses a sense of humour, and treats them as human beings. In other words, is not a boring "stuffed shirt" and stands up for them in their trials. This, I suppose, applies to all professions, but the sailor above all does not much like an obviously incompetent seaman in the broad sense of the word because, apart from anything else, it is frightening and dangerous in action and in severe weather. Can you blame him? I don't.

Much the same applies to going into action in destroyers and MTBs. As a passenger or underling one can tell very quickly if the captain has the touch, and if he has not it is very frightening—rather like being driven in a racing car by a very bad driver. This has little to do with courage, as such. Some have much too much courage and slay hundreds quite unnecessarily—and probably themselves. However, it is a necessary quality to have that second sense to know when one is engaging the enemy at the right time and according to a sound plan; and goodness how quickly one sees it as an onlooker when on the bridge. Without it one loses unless the enemy is worse!

Sadly enough, really I got terribly bored with the Navy owing, I suppose, mainly to a stint of at least 5 months up the Yangste off Nanking at the time of the war between the South Chinese 'Cantonese' and the North. It is not always appreciated that apart from considerable barbarities these people of the same race talk different languages they themselves cannot understand.

Most of that war was pretty dull, but being in the middle of the river Yangtse between both sides there was a good deal of sniping, especially when our troops had to go out on to the upper deck to walk forward to the "heads". The officers did not have this trouble. However, I do not believe anyone was seriously damaged; but it must have stirred up the bowels.

There was only one rather exciting occasion when I was ashore rather close to the time when the Cantonese were about to take the City and I eventually sheltered in the Consulate which was adjacent to the wall surrounding the City. For some reason I had on that day the use of a car which was certainly not mine. However, I got stuck by troop movements and the Consulate, apart from being some sort of shelter was being attacked and the consul was actually wounded. The marines guarding the establishment in many cases had their fingers cut off as a quick method of taking away a ring!

The situation had reached the point where it was most likely the Consulate would have to be evacuated to our ship HMS *Emerald* and I did what I could to organise an escape route, which involved descending the wall. This I eventually did by myself with a scaling ladder, and went to tell the Captain about the position in the Consulate, and to organise a cutting out party which was to be led by me in the night. Probably luckily, this never came off, though we prepared for it and eventually the Consulate people were allowed to come aboard by the normal route after a fairly heavy bombardment from both the U.S. destroyer and ourselves—a light cruiser.

To give an idea of the civilisation of these people at that time, the typewriters in the hulks used at that time as offices for most of the European firms trading in China had neatly placed excrement on top of them—quite a feat I would think.

While I was in the Navy I became owner, as I mentioned previously, of the Brescia Bugatti racing car. This beautiful little car, specially designed for the Isle of Man race, in which it was driven to victory by Pierre de Viscaya, became my property in 1923 while I was doing my engineering courses at Keyham and Greenwich. As the photo of her shows she was stripped down according to the fashion of the day for racing and hill climbing. She had a low axle ratio for hill climbing and acceleration on the flat.

One feature was the "blue" note made by the exhaust when about 4,000 r.p.m. was reached. The other peculiarity was the existence of two magnetos in the dash board driven by an extension from the camshaft.

I had a lot of fun with this little car and entered it more or less successfully at Brooklands and in hill climbs, notably Aston Clinton and South Harting. The cups won in these contests adorn our dining room table to this day.

I well remember Aston Clinton, which took place on the property of one of the Rothschilds, and as it was private property one of the stipulations was that we were not allowed to practice beforehand. The result was nearly disastrous in my case because I took one of the unseen corners towards the top too fast and went off the road into the spectators, who luckily melted away and no one was hurt. I still have a news cutting of this performance in which the author says that having left the road I returned to it, and finished apparently unperturbed at my experience. To the best of my recollection I was much perturbed. However, she was a safe little car to drive on the whole and I only remember one other rather alarming experience, which took place on a Good Friday as I was going pretty fast round the corner of a narrow Dartmoor lane with hedges on either side. Being late I was turning up the wick, and to my horror found a steam traction engine in the middle of the road. There was nothing I could do to avoid it other than to steer for the fence and land upside down in the field. Luckily the two men in the traction engine saw my predicament and lifted the car off me, but not before I had some bad moments listening to the dripping of petrol. Little harm was done and I was not much hurt, but it taught me a lesson not to drive fast on Good Friday.

2

Flying

Having left the Royal Navy in 1928 I joined the Auxiliary Air Force in 1929 after being taught to fly on a Renault Avro by Colonel Henderson, a well known fighter pilot at his school of flying at Brooklands. I gained a flying licence from the Air Ministry, No. 1539 dated 13 September 1928. This was what is called an "A" licence and I afterwards passed the "B" licence which enabled me to fly for hire or reward. I was also granted a Certificate of the Federation Aeronautique International dated 9 August 1928.

By joining the Auxiliary Air Force (later Royal Auxiliary Air Force) I was trained to a much higher degree of competence, first on the Lynx Avro, which was 100 per cent aerobatic and good for cross country flying. Our instructors at that time were recruited from the pick of Royal Air Force regulars.

My instructor was Flt. Lt. Dermot Boyle, afterwards Chief of the Air Staff, but at that time one of the leading aerobatic pilots in the Royal Air Force. As can well be imagined tuition from "Boylo", as we all called him, was of the best. The other assistant adjutant and flying instructor in my time in the Squadron was John Gillan, usually known as "Gillo". He was almost as good a pilot as Boyle but rather more wild in character and every-day life and probably took greater risks as a pilot. He was a Presbyterian parson's son from the Highlands of Scotland and was an amusing, and by no means devoutly religious character.

Both my wife and I, as well as Maime, my sister-in-law, took Gillo into our close friendship. At the outbreak of war he commanded a Hampden Squadron of bombers, which he managed with distinction, although by temperament he was more the fighter type. He stayed with us at our home, East Hoe, Hambledon, near Portsmouth, in about 1942 when posted for a short while to a Beaufort Squadron at Tangmere. The last we heard of him was that he had been given command of a single seat fighter squadron—I think Spitfires—which by this time had become the province of the very young and very quick on both sides. Commanding bombers, or even the two engined night fighters such as the Beaufort was not a good lead into the day fighter squadrons at this time. Gillo, good pilot that he was, was killed over the sea—probably the eastern end of the English Channel, and so far as I know he was never found. His last words on the R/T were "Damn, I am being shot down" as one of the latest Fokker Wolf fighters got on his Spitfire's tail. This was typical of his carefree attitude to life and its alternative.

In fact these instructors made it all seem so easy that one tended to become over confident to the extent that, when put on to the DH9a with Liberty engine, the light bomber with which the squadron was then equipped, I used to side slip the height off her when coming in to land and became remarkably unpopular with our very long suffering adjutant, Squadron Leader Bill Thornton, who was a regular and eventually became an Air Marshal and regular incumbent of the Whites club dining room, bar, and smoking room. Most of my fellow officers in 601 squadron were members of Whites.

I obtained my "wings" on the DH9a, which was an ex World War I light bomber and really obsolescent; its Liberty engine was a V12 designed and built in U.S.A. for wartime use and as engines went in those days it was good, but as can be seen from the picture the exhaust pipes discharged themselves each side of the cockpit and just ahead of the pilot who could see the smoke coming out in puffs as the engine was opened to full throttle for take off over the mass of houses surrounding Hendon airfield. However,

Fig 4 DH.9a Aircraft

Fig 5 Wapitis aircraft in formation

no Liberty engine ever cut out on me so that one cannot complain. The 9a was not very easy to land; at least I did not find it so. The undercarriage consisted of undamped rubber shock absorbers, so that the clumsy landing caused a series of ever larger bumps into the air accompanied by petrol slopped in one's face from the gravity tank located in the centre section of the upper wing just over the pilot. All one could usefully do in such a case was to go round and land again or give her enough engine to stop the bouncing and lift her sufficiently to make another 3 point landing—if you had room on the airfield.

The next operational aircraft with which we were supplied was the Westland Wapiti, which was an improvement on the DH9a, though it was alleged to have the same wings. The engine was the famous Bristol "Jupiter" air cooled machine. The undercarriage was much better damped and there was not the same tendency to bounce on landing. Although safe it was not at all easy to make a neat 3 point landing, as the elevators lost their sensitivity in the pitching axis near stalling speed. A little engine throttle towards the end of the glide would make a 3 point landing much easier by keeping a good air flow over the elevator surfaces just before the wings stalled.

When I thought about it at that time the obvious role for the Auxiliary Air Force seemed to me not day bombing but fighter duties (avions de chasse). Afterwards as the second war approached this was appreciated and the squadron distinguished itself in the Battle of Britain flying, mostly with Hawker Hurricanes.

I cannot claim the glory as I had left the Auxiliary Air Force long before this and had gone back to the Royal Naval Emergency List, after joining Vosper who were at Portsmouth. I was reluctantly compelled to resign from the Auxiliary Air Force as it was too difficult to work in Portsmouth and fly from Hendon in the week ends. I put in a few hours at Hamble on a reserve aircraft called the *Atlas* but derived little pleasure or benefit therefrom. One of the airmen who had prepared the *Atlas* for my first flight was encouraging enough to describe this aircraft as "The finest killer in all the Royal Air Force"—which strangely enough did not encourage me.

However, to revert to the 601 days, we went each year for two weeks training at Lympne aerodrome next door to the palatial abode of Sir Philip Sassoon, our Squadron Commander after Lord Edward Grosvenor, who was my first Commander. Ned Grosvenor was loved by all and when not commanding the squadron could usually be found in reasonable proximity to the bar at Whites. No more charming man could be met and it was a great sorrow to us all when he died at the early age of 40. He was, incidentally, an uncle of my wife's, which may have helped me to join this squadron, plus the fact that I was also a member of Whites, which seemed to be a good passport for selection.

Appreciating, even in 1930 that the correct role for the London auxiliary

Fig 6 A Gladiator

squadrons was fighter work our adjutant very sensibly arranged for visits to be paid to us at Hendon by fighter squadrons of the Royal Air Force, where we carried out manoeuvres together and, if lucky, were allowed to fly their single seat fighter aircraft.

One day I was sent up in my Wapiti to act as target for the fighters to make dummy attacks on, and managed to climb through about 4,000 feet of cloud by steering an easterly course and watching the bubble in the inclinometer as well as the air speed indicator—we had no blind flying instruments such as artificial horizons, turn indicators, etc. in those days. What one had to avoid was climbing on a northerly course as the magnetic compass was not reliable on this heading due to what was called northerly turning error. Having climbed out on top of this layer of clouds, I waited for the fighters to come up through the 4,000-feet layer and attack me, but no one appeared much to the annoyance of the Squadron Leader on the ground who was equipped with *Bulldogs*, a nice little fighter but also not equipped with blind flying instruments at that time.

As a treat for evading the fighters I was allowed to fly one of the *Bulldogs*, which was my first experience of a single seat fighter. At about this time I flew the *Siskin*, also a single seater. It was not until many years later that I found a more pleasant fighter in the *Gladiator* which was the best aerobatic aircraft I have ever known and fitted with all the aids such as a blind flying panel and brakes.

As those old enough will know, 601 Squadron fought with distinction in the Battle of Britain, including such stalwarts as Max Aitken, as he then was, and Archie Hope. I was proud to have been flying with them 10 years previously and to a large extent they taught me enough to take an insignificant, but interesting part in the Fleet Air Arm in 778 Squadron for a few months in 1940–1 while waiting for the U.S. engines to arrive for our Vosper MTBs. I also did a good deal of ferrying with 780 Squadron from Lee on Solent after going back to Vosper, as I was qualified and could help.

My days in 601 Squadron were happy ones, but I put in so much flying that my earning power as a civilian was much affected, thus when Glen Kidston asked me to help him at Vosper's Portsmouth shipyard for the princely sum of £500 per annum—worth about £2,500 today—I brought the family down there. We took a house at Bosham called Cut Mill belonging to the Tennyson Jesse family, at that time well known authors.

3

A New Kind of Life

In about April 1929 I left the Royal Navy, resigning from it just about when we were married.

I have never fancied married existence in the Royal Navy—most difficult to set up a home and very expensive to be constantly changing house, especially if there are little ones. The Admiralty let down engineers, or went back on their word, which I could not stomach having always been an executive officer till 1926 or thereabouts. But I have always been grateful for my time and training and managed to make many good and lasting friends. The training academically was first class.

Our first home was in London, a nice little house where I had some consulting work and did not really do too badly. But such business has its limitations unless one is very lucky indeed; it is surprising how many rich and otherwise pleasant people and firms resent being told the truth—mostly about inventions or processes into which they want to, or actually have put money. One of my more interesting jobs was in connection with a supercharger being developed by Captain George Eyston of record breaking fame—such a nice man and true friend.

Life in London, even then, was sometimes difficult and expensive, so we were both delighted to be asked to do a job down in Portsmouth for Commander Glen Kidston, who was building a motor boat at his own expense with the idea of improving the breed of boats in the Royal Navy, in which he had found many shortcomings while serving in aircraft carriers and other vessels. The firm he selected was Vosper & Co.

There was, of course, a certain amount to be said on both sides, but I have to admit that the motor boats supplied to H.M. ships for ship-to-shore work at that time left a good deal to be desired. I know, having spent years trying to keep them running as an engineer officer.

It is needless to go into the details of the shortcomings here; but those at the Admiralty at the time responsible for ordering and specifying the boats and engines demanded such difficult and unnecessary specifications that

Fig 7 The late Commander Glen Kidston

Fig 8 MV. *Advance*

only a few engine and boat builders were prepared to go to the trouble of trying to meet them. The intentions were good enough; but motor engines for marine purposes are difficult beasts to keep running unless built in large numbers and based on a unit which has been widely used ashore in the first instance both at home and abroad.

This, however, did not happen and a number of different types of boat and engine appeared which were the bane of an engineer officer's existence, and of all who had to run and repair them.

Commander Kidston's particular boat and engine had, I think, been started already when I came on the scene, though undoubtedly I had a certain amount of say about details and testing. The hull was designed by Fred Cooper and was being built at the yard of Vosper & Co. of Portsmouth, who had undertaken this sort of Admiralty work for many years. I well remember the Vosper name plate on my first steam picket boat when serving as a midshipman. The engine of Glen Kidston's boat was one by Thornycroft and the hull was of the hard chine planing type in which Fred Cooper specialised.

There was so much to do to prepare a craft of this sort, which had to be hoisted into a battleship and be suitable for naval service, that it became almost a full time job for me and Glen Kidston. He had left the Service just before me and was negotiating to purchase the control of this go-ahead little shipyard at this time. So he asked for me to be taken on the staff of the firm in a full time capacity.

At this time most of my experience had been as a naval officer, both executive and engineer, and I had a great deal to learn about the workings of a shipyard—even a small one. For sure I was an awful nuisance, but everyone including the managing director, Mr. Prangnell, was enthusiastic and helpful in the extreme. We completed the boat, which we named the *Advance*, and proceeded to demonstrate her to various Admiralty officials and naval officers, most of whom liked her; but the Admiralty, being what it was, showed little enthusiasm to buy the boat, though she did represent quite an advance on those in the Service.

Eventually we persuaded the Admiralty to allow us to loan her to HMS *Hood* for use during a cruise to the West Indies, on the basis that a report would be sent to the Admiralty on her general suitability for naval service; also on her performance and reliability.

Great excitement prevailed and in practically every way she behaved very well. We had a good report, with of course some recommendations for the future. And that was that—no more was heard. This was nothing to do with the ship's officers, who would have loved to have kept her; but it was, I suppose, too much of a bitter pill for the Admiralty boat section to accept that a naval officer could produce something better for the Navy than they were able to achieve with all their experience.

I don't doubt the cost was large compared with the existing motor boats, but there could be no real comparison, the specification and concept being quite different. At this stage Glen Kidston was unfortunately killed in an

Fig 9 Fast Motor Cruiser Silver Star

aircraft accident in South Africa.

At about this time we managed to get an order from one of the Wills family for a fast cruiser by the same hull designer, though we streamlined the deck structures above deck level at the owner's request. Here we had a twin Kermath installation—petrol driven—and for those days quite highly powered. The speed was 22 knots.

There was an appalling rush to get her built and finished in 3 months from the order, and she was shipped in a Dutch liner to Genoa, from where she proceeded to Cannes in time for the owner to have her for his holidays. I went out with an engine fitter and Lord Hardwicke, a director, and took her over at Genoa to run along the coast to Antibes in the first instance, so as to get her ready to run her round to Cannes the next day to meet the owner who came straight from the "train bleu" on board and to sea.

To the best of my recollection we were not really settled down properly as the clutches were very apt to slip and the whole engine installation wanted pretty careful handling. Also the electrical installation was somewhat overloaded, so I asked the owner not to run her at night until we had put one or two things right. All went well until my wife woke me up in our hotel room facing the sea at about midnight, and to my horror I heard the beat of the *Silver Star's* engines, and looking out of the window saw her crossing the Bay towards the Casino—most of the lights on and with a searchlight shining vertically. I was fairly worried and perhaps a bit angry, because this was almost bound to lead to trouble, especially owing to the shortage of electrics. At that time, also, it was not too easy to navigate in the darkness in that area as there were unlit islands and rocks so it was undoubtedly difficult to know exactly where they were.

However, I rushed over to the Casino in my pyjamas, fearing the worst; only to arrive in time to see them getting away towards the islands, which I knew to be a pretty hazardous piece of navigation, though easier nowadays.

I went to bed still fearing the worst, and found next morning that their boat had run at full speed over a tongue of rock sticking out from one of the islands and, as can be imagined, the stern gear and propellers were in a poor state. However, it could have been much worse and gave us a day or two of respite to get things straight. The next drama occurred when she went on fire for some reason on the way over to Monte Carlo one evening. This finished the season; and by this time I was in a nursing home at Cannes with a boil in the ear (otitis media) caused by diving under the boat to see what was wrong with the stern gear. My ear had to be operated on.

Then the next incident was that they managed to drop her from about 40 feet when hoisting her out of the liner at Southampton on return, so with the insurance money we did a pretty major alteration eliminating the vee drives because we thought their additional inertia was overloading the clutches.

This was quite a satisfactory fast cruiser for many years, and I heard of her during the war, but not after. We had learned a lot. Still I meet the eventual owner Jack Dunfee, who certainly enjoyed his week ends in her.

At about this time, owing to what is usually called a board room row I was asked to take on managing director of Vosper & Co., as it was then; we were in fact owned, so far as the majority share holding was concerned, by the local coal merchants, Fraser & White, who were next door to us on the Portsmouth inner camber. The chairman, Harry Gilbert, who was also chairman of the Vosper board, persuaded me, after some reluctance on my part, that I could do the job and I must say he was always a great support to me. I certainly learnt a lot about business from him, and about local government as well, as he was one of the most influential Aldermen of the City of Portsmouth. This was July 1931.

I think he had bought up Vosper only a few years back to avoid it getting into the wrong hands or rivals of his as it was on the inner Portsmouth Camber. It had proved rather a hot potato, as at that time (1930–1) we had little work and only a total staff and workpeople of about 70. However, I had little to lose and took it on, remaining managing director for 32 years—not all easy ones! I don't think my appointment was universally popular; in fact I am sure it was not, and I was very conscious of the fact that my training to date had been that of a naval officer, and I was well aware of what would be said on this subject if I let the firm go bust.

However, I cannot say that with very few exceptions I was not enthusiastically and well supported throughout the more difficult, and in fact all, times. But let nobody think it is an easy job to build up a boat building and engineering firm like this and make it pay. There are very few in this cate-

Fig 10 Jolly Boat *Handy Andy*

gory. There is difficulty in naming those who gave me the greatest support for fear of seeming unfair to others.

It is probably fair to say that with a good background of repair work and engineering, together with quite a lot of yacht repair and laying-up, what really got us going was a couple of Ford engine conversions to marine purposes, which enabled us to become competitive with our rivals in the fast boat world—particularly with The British Power Boat Co., owned and run by Hubert Scott-Paine, a flamboyant and persuasive character who had bought it with money he received from his share in the Supermarine Aircraft Company.

The Ford V8 of the day, giving about 75 BHP (max.) was probably the most important of the two, although the Ford 4 cylinder "B" was the first to bring us into the limelight in a small 16 foot planing dinghy we designed for the Royal Navy, which we called a "Jolly Boat". It is not possible to ignore the outstanding work done by Fred Street in connection with these engine conversions. He later became our Engineer Works Manager for many years and has only recently retired.

My efforts as a managing director in those far distant days, the early Thirties, left much to be desired, but one was faced with a big problem, despite everyone's cooperation. A managing director, even nowadays, with the many experts, consultants, management techniques, courses and so on, has really to review the overall picture and decide on which of many problems he must concentrate, leaving others to take on problems which perhaps suit their particular skills better. Also, once the company expands beyond a certain size the managing director's job becomes quite different and in many ways bigger for obvious reasons.

At the stage I am talking about, besides getting the management and personnel right, which was absolutely a *sine qua non* to success in difficult circumstances, I spent most of my energies on the technical side besides on selling and getting contacts in the right places. Rightly or wrongly, and I hope rightly, I soon came to the conclusion one must learn more about the design of boats oneself; for in the long term one gets the blame if anything is wrong and the overall design, including performance, behaviour, and appearance make all the difference to clinching a sale.

Equally, I was finding myself testing and clearing what were really prototype troubles in most cases, a matter not by any means always easy in small high speed craft. This applies to both hull, machinery and controls. Trim is the all important factor. Here I was much helped by my flying background which had been reasonably extensive over the years. There are many analogies between handling a fast craft and an aircraft—probably more than most realise. Flying teaches you an attitude of thought in dealing with matters of behaviour where trim and dynamic problems generally, including control surfaces, etc. are all important.

Anyway, though brought up as a marine and mechanical engineer with a bit of aeronautical engineering thrown in, I took on the responsibility for the design of our fast craft and yachts in about 1934, and retaining it up to about 1967, where Vosper had the design responsibility. This does not by any means infer that alone I did it. Anyone who tried to do this would have come unstuck and have had no time for the more important matters; but I used to give sketches or drawings sufficient for the draughtsmen to get on with the job. For instance, to draw out the lines plan for a fast planing boat, or any boat for that matter, is a highly skilled exercise, and though I would draw to scale a few of the more important lines such as chine, keel and gunwale, also deck structures, I would leave the production of the detailed and faired plan to someone much better at it than I. The final fairing in any case has to be done in the mould loft and on the boat.

This seemed to work all right as long as the draughtsman did not try to go too far away on his own; and it is, I suppose, the way most naval architects get started on the design of a ship or boat. Design at this stage is a very tentative process which cannot be rushed, and it is usual to have to make several shots at achieving the optimum until submission to the owner and commencement of production. There is little really easy about it, at least not if one wants to live with the world in this sphere of activity.

I seldom cease to press the point with both our own people and customers that in the case of a prototype fast boat it is remarkably seldom that, with the best will in the world, one does not have a fair amount of fiddling about and alterations in the initial running stages. Small changes can make an immense difference to all round behaviour and performance, and until the craft has been run it is virtually impossible to anticipate these alterations.

The job is akin to the production and testing of prototype aircraft, only considerably simpler and less dangerous. It is my experience that on the whole the trials officer or test pilot for this type of craft is best if he has had preliminary experience with aircraft, and high performance ones at that, a little of which I had in 601 Squadron A.A.F.

Most of what I have discussed up to date relates to technical matters but it was obviously necessary to administer the company, as by this time I had the overall responsibility and it was up to me to make it prosper—no easy task.

I still felt the number one priority was to get work and good work at that. Having got it, the job had to be executed as well as anyone in the world could do it. It will be realised this was no particularly easy task; because, though the British—contrary to what people like to say in this day and age—workmen are conscientious, their trouble is being difficult and jealous. For one like me, fairly new to the job, it was mighty difficult to get them to do what I wanted, which might not have been wonderful or even right.

In this process many, mostly those in the higher echelons, had to be eliminated where necessary, and I suppose this was one of the worst jobs, which everyone loathes. It is, however, essential to success, the quicker the better. In those days there were no consultants available to advise the management who to get rid of, so the responsibility and unpleasant task fell fairly and squarely on my shoulders. I think few mistakes were made in this department if one looks back over the years.

Harry Gilbert as chairman was a tower of strength and he would always support me through thick and thin. Some of his co-directors who were on our Board at this time were not at all like him. They were not always helpful; which is understandable, as the decisions were almost entirely technical, which they could not understand. The exceptions to this were Rad Guard (O. R. Guard) and Philip Hardwicke (The Earl of Hardwicke), who were very helpful in our early days.

It will be understood that until about this time, Harry Gilbert owned the whole of the equity of Vosper & Co. and in fact was absolute boss. His philosophy of what may be termed "business in the large" was fascinating and obviously successful. Few, if any, recently retired naval officers would have any idea of this department, but it is all important. Until you understand the workings of the minds around you there is little hope of success in the commercial world. I always remember one of his favourite pieces of advice: "Remember it is easier to catch a fly with treacle than with vinegar!" How true!

If you go about it the right way and make friends with your rivals they will even tell you how to get on with captains, engineers and stewards, which that great hero in the world of naval architecture, Charlie Nicholson, gave to me as advice not long after I had taken over this new job at Vosper, but not before I had started to take some good and profitable work away from him, whose yard was the other side of the harbour. This piece of real help to a young rival I shall never forget—a truly big and generous character—as well as a brilliant international designer, mostly of sail but of some fine powered yachts as well. Incidentally, he was of an earlier generation to me, and it is nice to see his firm still operating brilliantly some 36 years later. His charm has been passed on to the other members of the family and his grandchildren.

Of course, while a pleasing personality and sincere desire to help the customer is an essential, there are obvious cases where it is necessary to be really tough, when a customer is impossible to deal with. One is lucky if one can sense this before becoming involved, and it is an important part of a manager's duty. In fact I have always said that such success as we had in building up the firm was due to an appreciation of the fact that the success of our company is *even more due to the contracts we turned down than those we took on and completed.* There are many potential jobs which can ruin a firm, and too many will do this if one is not very careful. These will also have the effect of making it impossible to take on the good ones when they are available.

Any firm of our type is inevitably limited in the number of effective people who can be put on to the preparation of estimates, design, specifications, and vital preliminary work; and it is fatal to take on too many contracts and make a poor job of all of them. Much good judgement and firm decision is required here, and it is not by any means easy to select the right ones. This still applies.

While on the subject of the business or administrative side of running a company, it has to be said that though unethical it is seldom that one gets a decent job without giving to one or more what is, perhaps, best called a "backhander". This is probably wrong but practically always necessary if you don't want unnecessary trouble. Anyone will know this, though they may not talk about it. The art is to know whose hands to put it into and how much! Some individuals are quite open and tell you beforehand what they want out of the job, which helps a lot. I fear it is not unknown in business for a manager who gets close to his opposite number in the purchasing firm to add on a bit for himself with the connivance of the other party, and get it back when the bill is paid. I have not come across many of these, but they cannot be kept on the pay roll and it is almost certain they will be given away by someone who gets to know eventually. One good sport, who for a time was the mate to an owner-skipper who gave us a lot of work to do, was a bit of a "Bible thumper" and gave all our backhanders straight back to his owner, my very good friend. This caused much amusement some years later, when I was told about it, and his wife, a young and beautiful girl, thought I was an awful crook and told me so! The owner concerned, Loel Guinness, one of my best friends, did not seem to mind too much as he gave us at least two of his conversions from naval craft to do after the war and also a lot of work in his last and biggest yacht.

This is enough on the subject of the frailty of human nature, but I must add that, to my knowledge, no one in the armed forces, particularly the British Royal Navy, ever looked like taking a bribe. A civil servant, in the early days laid himself open to suspicion, but I never had any positive knowledge of this and it, perhaps, should be added he was a civilian who had really been too long in one office and had probably got too thick with some firms which caused others to chatter. There was, however, I admit a suspicion, and he behaved unwisely to say the least.

In a number of foreign countries this sort of thing is practically a way of life and there is little hesitation in saying exactly what is wanted and by whom, and the art of the negotiator is to pay out in the right quarters sufficiently to get the job; while being absolutely mum as to what is happening even within his own firm.

By this time, which would, perhaps, be 4 years after becoming managing director, it became obvious that the Royal Navy were ripe to take on some fast boats incorporating new ideas both for ships tenders, besides actual small fast warships.

Not all were too keen on this, but as the Controller of the day was in favour, it was pushed through; and I suppose it is fair to say that Hubert Scott-Paine, our rival over on Southampton Water, played a big part in having encouraged the Royal Air Force rescue service in the first instance. These craft seemed to be giving satisfaction in the service of the R.A.F.

It is probably true to say we could have designed and produced the hulls to satisfaction at this time, but were pretty well stuck for British engines, and as a matter of shrewd business Scott-Paine would not sell us his Meadows marine engine.

Lawrence of Arabia, who was at that time an aircraftman, took a considerable part in the development of high speed rescue craft and loved being in a boat. I had the pleasure of meeting him a few times and interesting arguments about fast boats invariably followed.

However, by making friends with Ford Motors at Dagenham, and working at the necessary marine conversion, it was not long before we were knocking at the door of the Admiralty and R.A.F., and pretty soon we had to be let in, much to the annoyance of some. There was some pretty hot competition, and almost unbelievable lengths were gone to in order to beat us.

One day the steward of a well known yacht club near Southampton came to tell me a high official of some other company was wanting to send over what can only be termed spies to get into our yard and pay some of our men for acts of sabotage, such as putting sand in bearings. I was not able to verify this story, but the steward, who was ex-Navy, was convincing. I believed him and took the necessary steps.

The pace really hotted up when we produced a motor torpedo boat, as they were then called which, fitted with three Italian Isotta Fraschini engines and two 21 inch torpedo tubes, it was a very much faster and more potential weapon than the smaller boats then being pressed on the Royal Navy by our rivals.

For a long time we were kept at arms length by the authorities, but as we had paid for it ourselves, and the weaponry in the form of torpedo tubes and firing gear had been designed by HMS *Vernon*, the torpedo establishment, it became impossible not to give us at least a trial. The sticky part, so far as I was concerned, was the necessity to demonstrate the seakeeping ability of our craft in a Force 7 gale outside the Isle of Wight, which I had to do myself in control of the boat.

This is a pretty formidable test for a 68 foot, 40 knot boat, and I was not too happy during the period we had to wait. The great day came and they sent a destroyer out with us. Though we got through the test fairly well, I made the mistake of driving from inside the wheelhouse from where, despite a so called clear-view screen, I could really see very little. I have since found windscreen wipers much more satisfactory as do airline pilots. The type fitted in the *Dimarcha* cannot be better.

Afterwards one of the spectators in the destroyer congratulated me on a wonderful piece of showmanship in cutting across the bows of the destroyer. I had to explain that I had no sort of idea this had occurred and that it was quite unintentional. However, we got away with it and looking back it was really asking a bit much of a 68 foot boat, and later trials were moderated. Needless to say in this seaway one could not maintain full speed. This spectator took a wonderful cine film which like a fool I lost.

There could really be no question of fighting a boat in these conditions against any formidable enemy. Get back to harbour, yes, but not fight, which is the object of the drill. Subsequently most, if not all, of these craft have been fought from an open bridge. Recently nuclear warfare has necessitated closed, air conditioned bridges, but we also have a radar plot below.

Fig 11 MTB 102

This hurdle more or less over, we then had to attack or importune the Admiralty again and after some time she was bought from us at a cost of about £16,500 so far as I remember. She was named *MTB102*. Today the cost would be not less than £400,000, probably more, though quite a lot of this increase would be due to additional complication, weapons, etc.

As Scott-Paine would not sell us his engines to fit into a planing dinghy (not that I blame him), we had to find a marine engine for ourselves. There were plenty available in the U.S.A. which worked very well, but we could not really get down to a workable English marine engine having the right power to weight ratio. Eventually we found a charming and very well known English autmobile firm who took on at least the major part of the job of conversion to a marine engine, and the Admiralty agreed to allow the resulting *Jolly Boat* to be tried as tender to one of the Royal Navy cruisers of the time. Frankly, the boat was not a success for various reasons, the chief of which was the fact that the power delivered by the engine was insufficient and was of the wrong characteristics for marine work, which really requires the short stroke, large bore, engine whereas British automobile engines to get their power and keep down the rated horse power, had to have long stroke, small bore, and high revs.

This led to the boat barely planing, especially when there were too many aboard and resulted in a very wet boat. Something had to be done to save the day and at my suggestion our engineers converted a Ford industrial 4–49 which could be bought by anyone, and we also took an American licence for a marine gear box, called the Joe's gear made by Snow & Petrelli. The result was successful and when the cruiser, at our request, sent the boat in for overhaul we changed engines—at our expense of course—and returned a really useful boat.

There was some speculation when this was realised in the cruiser; but as the boat was on loan and belonged to us they left the matter alone, having no real responsibility other than to make a report; so the cruiser took the boat back and ran it with great success. This was, of course, naughty of me but we would have been right out on our ears had we not done something pretty drastic about it all.

As it happens it was a success and I well remember, and so does my long suffering wife, an occasion when the Admiralty agreed to give it an official trial with a view to making this boat a standard article for service in the Royal Navy.

As it happened, most unfortunately, the day before the trial we managed to run the main bearings, so things looked mighty black. However, our engineers worked like slaves and we got her running again by about nightfall. She had to be run in all night, mostly in Portsmouth Harbour, and my wife and I took our 3 hour turns along with the others, including Harry Cole. I well remember the 4 hour full power trial of the following day, which was completed without a hitch. We could quite easily have been kept out of the Royal Navy for many years, if not for good, if some pretty smart operations had not been laid on to get her going in time. Subsequently we changed the hull somewhat, using a clinker construction and modified form. This turned out to be the first actual hull I claimed to have designed, and it was run for many years as a small fast work boat for our yard called the *Handy Andy* (see Fig. 10).

One hundred and thirty of these little boats were produced. The first large batch to be ordered by the Royal Navy was sent to a new destroyer flotilla in time for the Jubilee Review of King George V. Our rivals, which I think were called safety dinghies, were supplied to the previous flotilla and of course there was great competition between the two flotillas to keep their boats running reliably. Our consternation can well be imagined when we heard of a signal sent from the flagship, to whom one of our Jolly Boats had been sent for duties as despatch boat for the day, saying the boat was unreliable and demanding another boat. We were subsequently informed by the flotilla engineer officer that the fuel tank had been found to be full of bird feathers! Further examination showed that all our boats had been similarly treated, so it was concluded that there had been dirty work at the crossroads from some source.

These craft and the anecdotes surrounding them are not entirely described in chronological order, because the Jolly Boat incidents took place well before the trials and purchase of *MTB102* for naval service. However, if we assume most of this development to have taken place between 1935 and 1937 we should not be far wrong. In the meantime with the development of the Ford V8 as a marine engine, also using the Joe's clutch and reverse gear box, we were able to offer a number of interesting fast boats for naval and other service. Amongst these was a special 3-engined planing boat for use on board HMS *Hood*, as well as a number of 25 and 35 foot standard ship's boats. Amongst these were some rather smart admiral's barges. We built

Fig 12 Fast Motor Boat for HMS *Hood*

Fig 13 35-ft Fast Motor Boat for Royal Navy

Fig 14 45-ft Picket Boat for Royal Navy

two 45 foot picket boats, which were allocated to battleships, with no less than four V8 engines installed. One of these two, of which I still retain some photos, was used by me to make a run down to Ince Castle near Plymouth during the Easter holidays to visit Bobby Somerset, a well known ocean racing yachtsman who was at this time working for us as sales manager. Along with Bobby and my wife, Victory, I took Lord Hardwicke, who had been a partner of mine in Vosper since we had jointly bought the equity from Harry Gilbert. Actually my share was two-thirds to his one third, but between us we had the control of the company through the voting power of the deferred shares.

This is an arrangement which sounds fine, but it can and did cause difficulties at a later stage in our career. However, at this point in time when all was going almost unbelievably well, neither of us could see the black cloud over the horizon, which spoilt our friendship for a time.

The picket boat went very well in a strong following wind and sea blowing at least Force 6. We made a rather poor landfall after crossing Lyme Bay, having chosen a course which took us nearer to Berry Head, which in any case one tends to sight before the Start; this gave us a good opportunity to see how she behaved when running with wind abeam. For a short while, just to see how she behaved we put her on to a course heading into the wind, which gave us a good idea of her all round capabilities in a seaway. This is the sort of trial we always tried to organise with every new type. The picket boat was surprisingly good, as is always the case with a reasonably well designed hard chine planing boat. A speed of 20 knots was about her best cruising speed. After the Easter holiday Bobby and I took her back alone without even a mechanic to watch the engines. On the way down we had one of our young engineers with us, but let him go back home on arrival at Plymouth, as he was seasick and the engines ran well.

A matter of some interest arose in the trial stage of *MTB102*. Lord Louis Mountbatten had been much impressed by the potential of the Oerlikon 20 mm machine gun of Swiss origin and had been trying without success to get the Royal Navy to give it a trial. As our boat was on trial and not yet bought by the Admiralty there was nothing to prevent our fitting two of these guns for trial. Lord Louis suggested this to me and I fell in with the idea which forced the gunnery school at Whale Island to give them a trial. The guns seemed very successful and of course later in the war were bought from the Swiss factory and used in large numbers. Lord Louis can certainly be claimed to have shown foresight and good judgement as he did over so many other naval matters. Incidentally when we handed the boat over to the Royal Navy, the Oerlikon guns were removed, which shows the conservatism of some Admiralty departments at that time.

4

Record Breakers

Early in 1938 I was asked by Sir Malcolm Campbell to take over the design and building of a boat to beat his then existing world's water speed record which, in *Bluebird*, stood at about 130 m.p.h. Sir Malcolm and Reid Railton, who had a great deal to do with the first *Bluebird*, as well as a number of the world's record breaking motor cars of the era, wanted to design this record *Bluebird* on the principles of what is known as "three pointer", at that time being developed by Arno Apel in U.S.A.

Fig 15 Record Breaking *Bluebird II*

I was told by Malcolm and Reid that they had approached Apel direct with a view to his designing a hull for them to take the existing Rolls Royce "R" engine, at that time fitted in *Bluebird.* So far as I am aware Apel refused the invitation, though I did not see the correspondence.

It was the first time I had attempted anything in the world record breaking class, and it gave me food for thought; with the advice and assistance of Dr. Gawn at the Haslar Experiment Tank, as well as the availability of the Rolls "R" engine and the transmission arrangements already designed by Railton and fitted in *Bluebird,* it seemed a reasonable bet. So we took it on.

It was decided to start afresh and after a few shots we produced a model that ran extremely well in every way in the tank. It was, admittedly, using Apel's principle of the three pointer; but was not by any means a Chinese copy and in any case no boat in the U.S.A. had ever reached the sort of speed we wanted. I say this because the general configuration we were aiming for had a hollow tunnel running through the length of the boat separating the two forward or outside planing points (see Fig. 15). It did not take much imagination to envisage quite a nasty or dangerous aerodynamic lift problem due to this tunnel; so we arranged to test a larger scale model in the Vickers wind tunnel at Weybridge. They recommended the tumescence or hump, running through the length of the boat from forward to the engine casing, including one or two other lift spoiling arrangements. This can be seen in any good picture of the boat.

The propeller revolved at 9,000 revolutions, which is, of course, very high; but there were many advantages; not least being reduction of torque effect which would give the boat a permanent list at speed unless taken care of; also by virtue of high revolutions and therefore low torque the shaft and other drag forming appendages could be made much smaller. Dr. Gawn made himself responsible for the propeller design which turned out to be just about "spot on" for the speed we were predicting.

While in the process of designing the second *Bluebird* Sir Malcolm asked me to go out to Lake Geneva and subsequently Lake Halwill in Switzerland to watch his attempts to put up the record in *Bluebird* before running the *Bluebird II* the following year.

It was all very interesting to me, especially as Malcolm seemed to be having a good deal of trouble in controlling the boat in the higher speeds. He kept on asking me what was the trouble, and I found it impossible to give him much idea about this unless he let me drive the boat myself, which is far the best way to form a judgement as to handling qualities. At 4 o'clock in the morning of the last day we were out there, he suddenly told me to jump in and drive her as fast as I could, which was honestly not as fast as she could go. But I did reach 120 m.p.h., or thereabouts, when a certain pattern of behaviour commenced which allowed me to arrive at an idea of Malcolm's difficulties above this speed.

The features which I considered were causing this trouble would not in any case be repeated on the *Bluebird II*, so that further discussion and investigation would not be necessary. At Lake Halwill, Malcolm subse-

Fig 16 The author driving the first *Bluebird* on Lake Geneva July 1938

quently beat his own record about one month later. He had moved over to this very attractive spot near Zurich, I think because there was too much traffic on Lake Geneva, causing disturbances in the surface of the lake. Either this was the cause or there were submerged springs, which were alleged to produce disturbance.

The following months saw the *Bluebird II* taking shape in our Portsmouth shipyard and we were just about ready to run by the beginning of August. This time Lake Coniston in Lancashire was chosen and it turned out to be a very lovely and suitable place.

Bluebird II was put in the water for the first time on a Monday and by the following Saturday the record had been put up to 141·7 m.p.h. I don't really know how serious he was, but when I had reappeared at Coniston to ask how she was behaving on the Friday, Malcolm had taken me aside from the press and various officials to tell me she was awful and that he did not think he could even do 80 m.p.h. in her! This shook me a bit, and as he could not describe her misbehaviour at all clearly I could only suggest he let me have a run in her early the following morning. After consideration he said he would have one more shot first of all the next morning, when he immediately broke the record by a substantial margin unofficially, of course, as the official time keepers were not up there yet. However they appeared that evening and the record was confirmed at 141·7 m.p.h. the following morning. By this time it was obvious that war was imminent; so that some years elapsed before the record breaking game recommenced in 1946. But as we are on the subject of the world's water speed record we might as well continue, and recount what happened to me after a lapse of the war years, which I describe in the next chapters.

I do not doubt that by a certain amount of tuning in every way—engine, hull, trim, propeller—we could have reached 5 to 10 m.p.h. more with the *Bluebird II*. She was a beauty for behaviour; so much so that I did not get a ride at this stage! There was no need for it. Malcolm, however, had in the meantime got in touch with Frank Halford, the famous engine designer of de Havilland responsible for the Gipsy engine and other powerful ones used in the war. These were piston engines and he had subsequently produced a series of gas turbine reaction or jet units for aircraft. Malcolm was after one of these for the *Bluebird II* and I was brought into the picture again.

The whole arrangement of propulsion by the jet is different from that by propeller for various reasons, such as line of thrust, slipstream on rudder, and other matters. However, we did a good deal of investigation, again at Haslar under Dr. Gawn, and this time we reproduced the jet as best we could by using a Jetex rocket such as are fitted in model aircraft. It seemed to be a sound approach as we could reproduce the thrust and its line through the boat with very reasonable accuracy.

To give an idea of the effectiveness of this Jetex rocket unit: I bought for my children one of the model aircraft being sold on the open market at that time and the very first time I released it from the front of our house at Hambledon the machine soared into the air and turned east over a large belt of trees and was never seen again. It probably landed in the neighbourhood of Chichester some 15 miles away.

We had the scale model boat running very well in the large tank at Haslar and all seemed well. We got on with the installation in the full scale *Bluebird II* and had great hopes of increasing the piston engine record.

This, however, was not to be, owing mainly to the fact that at or just beyond a speed of about 90 m.p.h., a violent porpoising action was set up which effectively inhibited further increase in speed.

More models were run and still the forecast at full scale was good. Malcolm allowed me to drive her but the result was just the same; a characteristic porpoise, which is really a combined pitching and heave at a more or less steady frequency. There was a possibility we could have run through this range and become steady, but I came to the conclusion we could not.

I do not have any doubt that this could have been corrected, because the basic cause was the difference in thrust line; but it would have required some alteration to the relative angle of attack to the flow between forward and aft planing surfaces. This was, in fact, what we intended doing next when matters came to a standstill owing, so far as I know, to the fact that Malcolm wanted to take the boat abroad to some lakes where they had invited him to appear during the winter months. As the boat was by no means running properly at this stage, de Havillands, the engine makers, would have none of it and took their jet engine away, which was only on loan.

This finished the effort, much to my disappointment, because I was sure we could get her running given a little more time and the chance to make the above alterations.

It does, however, show that in some respects, especially in relation to behaviour, it is remarkably difficult, if not impossible, to make a small model behave in all respects as will the full scale job. We have found it on numerous subsequent occasions, even when we went to the length of making a special model of celluloid throughout in order to be certain that the moment of inertia about the pitching axis and radius of gyration were reproduced to scale.

The probable answer, which we had ready to try in the following year, was to alter the relationship of the angle of attack between the forward double planing surfaces and the aft surface. In the light of our later experience with a considerable number of fast boats this would have done the trick.

All of us became sadder and wiser people and although I was certain it was only a matter of time, poor Malcolm, who was not in any real sense technical, was dreadfully frustrated and never really recovered from the setback. He told me he had very high blood pressure and died a few months later.

I will not go into detail about the further running of the *Bluebird II* by Malcolm's son Donald, who inherited her; I frankly had little to do with the job.

I was next involved in record breaking with John Cobb. Unhappily this effort in the *Crusader* ended badly, but despite the tragedy at losing such a wonderful man, who already had the world's record in a Railton designed car, there may be some justification for describing some of the more interesting problems met with in designing this potentially very fast boat with which, at one stage, it seemed quite certain he would take the existing world's record, then standing at something over 160 m.p.h. This speed Cobb had exceeded many times on Loch Ness before the fatal day, though, unfortunately, without the official time keepers being present.

It is, perhaps, appropriate to describe the various design stages in detail as presented in an article I wrote at the time. It appeared in the "Motor Boat & Yachting" of August 1952:

"CRUSADER"

The Design and Construction of John Cobb's Jet-Propelled Boat

It was announced on 1 July 1952, that Mr. John Cobb was building a boat with the object of carrying out research into high speed on water. Mr. Cobb stated: "Its rather revolutionary layout was suggested to me by Reid Railton. The design and construction have been carried out by Vosper Ltd."

Behind this statement has been nearly three years of more or less patient research.

Reid Railton is one of those gifted engineers, who has that "second sense" about anything to do with high speed in any form, yet who can back up and confirm his "second sense" with an adequate use of calculation methods where applicable. We are indeed privileged to be associated with him.

In April 1949, Reid Railton, who lives in California, wrote to me a letter which really outlines the commencement of the project. The important part of the letter reads as follows:—

"I think it would be foolish for us to embark today (1949) on any project aimed at less than 200 m.p.h., which would mean designing for aerodynamic safety up to at least 250 m.p.h. If you agree on this it seems inevitable that our first line of approach must be to think up some "body shape" which would combine the necessary aerodynamic properties with adequate accommodation for the man and machinery and then (and only then) to try to work in the necessary buoyancy, planing surfaces, etc.

Fig 17 Original three-legged model

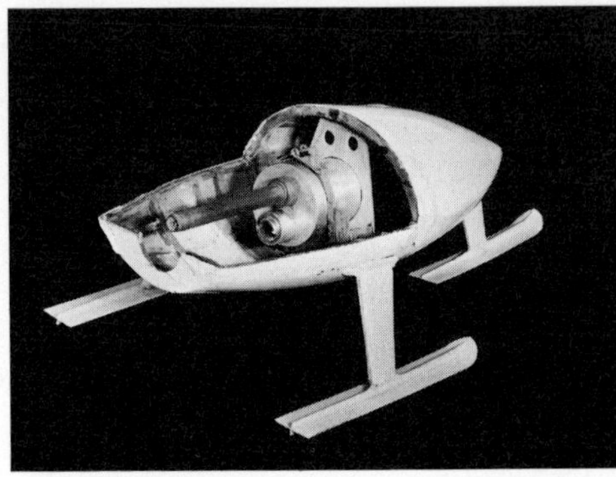

Fig 18 Original three-legged model showing rocket

The sort of thing I have in mind as I write is a very small seaplane hull with a couple of small floats or skis mounted on outriggers, where the tail normally is, i.e. a tricycle with the one wheel in front."

It should perhaps be mentioned at this stage, that the magnitude of the aerodynamic lift with increasing speeds at the forward end of the then conventional type of 3-point hull such as was exemplified by the *Bluebird* and many American fast boats designed on the principles originally propounded by Apel, had for long been the obvious limitation to further substantial development of this type.

In the course of endeavouring to translate Railton's ideas into a practical job I first schemed a layout incorporating an obviously good though somewhat impractical aerodynamic form of sufficient size to house pilot, engine and fuel. The aim here was to produce an aerodynamic form so that when the craft was running steadily at the designed trim there would be no tendency either to lift or depress.

Furthermore in the event that a disturbance such as a wave or "swell" causes the whole craft to assume an angle likely to cause lift it is then desirable to select a shape so that the resultant of forces due to lift and drag acts in such a direction as to restore the running trim angle rather than increase it.

On to this nacelle or hull were attached four legs with aluminium skis attached to the lower extremities, in such a manner that their angle of incidence could be varied, as well as the spread and span. The fourth leg was located aft with the object of damping out any inherent tendency to "porpoise".

The valuable advice of Mr. Gawn, superintendent of the Haslar Experiment Tank and his staff was available throughout and is hereby gratefully acknowledged.

The immediate object here was to

Fig 19

Fig 20

produce a system of planing surfaces which appeared likely to be able to avoid the "porpoising" at high speeds which bedevilled the *Bluebird* when a jet unit was installed.

In my opinion this trouble was not aerodynamic in origin as the forces due to air flow at the low speeds at which "porpoising" occurred in the *Bluebird* would be relatively insignificant, also incidentally the *Bluebird* had run previously at much higher speeds without trace of "porpoising".

While on the subject of "porpoising", there has for long been a convention in the world of flying boat designers to the effect that amongst other factors tending to eliminate this trouble is the incorporation of straight buttock lines for about 10 per cent of the forebody length ahead of the "step".

There does not seem to be any scientific explanation of this phenomenon and I am not convinced that it is in fact of overriding importance in our particular case, although it cannot by any means be overlooked. It so happens that in our final form as selected, where the original ski forms and angles have been used, the buttock lines are in effect straight.

It is hoped, and there is some justification for thinking, that the arrangement of planing surfaces we have now produced will avoid this instability by virtue of the relatively large longitudinal "spread" between the surfaces and to

Fig 21 One-sixth full scale model

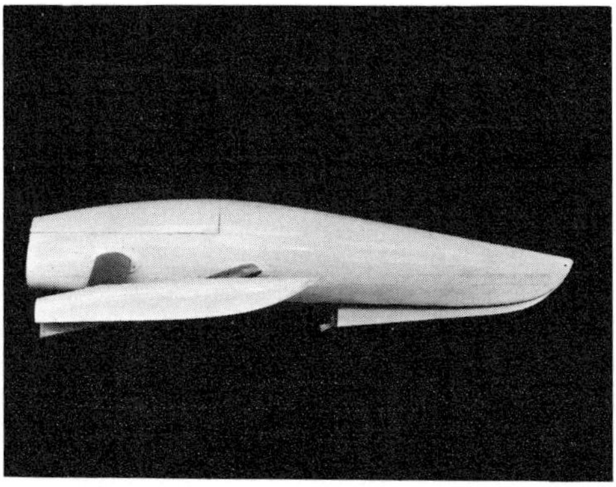

Fig 22 One-sixth full scale model showing 32lb rocket

some extent by very positive loading of the forward plane, brought about by a substantial differential between the "value" of the forward and aft planing surfaces. Whether this is borne out in practice remains to be seen, but this particular trouble has not shown itself at any stage during model tests.

It has to be appreciated that in this case we were faced with the evolution of a new principle or layout as well as producing from the result of this evolution a craft capable of speeds well beyond anything achieved up to date.

After some fairly considerable running of this model a combination of angles and other dimensions emerged which appeared likely to give stable running conditions at speed. This latter consideration was thought to be of more importance at this stage than was the achievement of maximum propulsive efficiency. In the course of this process the fourth leg was found to be redundant.

In the case of the three-legged model an air rudder or fin was selected to assist in keeping the free model straight. This was because only small water fins could be incorporated on to the skis. With the final model it is hoped to maintain directional stability by the fitting of relatively large water fins on the inside of each sponson aft, while control will be effected by a small rudder fitted abaft the forward planing surface. This is a somewhat unusual arrangement, but in this case has the advantage of simplifying the control linkage between wheel and tiller arm.

As this first model was one-sixteenth full size the required speed to reproduce 250 m.p.h. in the full scale was just over 60 m.p.h., which is more than can be achieved by the carriage of the "Experiment Tank".

In order to continue the experiments at one-sixteenth full scale and to reproduce "behaviour" at these speeds, it became necessary to develop rockets of approximately one pound thrust to propel the model along the appropriate line of thrust.

Messrs. Wilmot, Mansour and Co. of Totton, Southampton, undertook this work for us and there is no doubt that the rockets were quite invaluable in the work of testing for behaviour. Our first effort on a free rocket propelled model was as shown in Figs. 17 and 18 and the results were remarkably promising. The obvious limitation here was the necessity to launch from shore by running down sliding ways.

Some wind tunnel experiments were kindly undertaken for us at this stage by Mr. Hooper of the Fairey Wind Tunnel, and from that time onwards our efforts were concentrated upon producing the excellent running qualities of the ski or three-legged model in combination with a hull shape acceptable from an aerodynamic point of view, which yet possessed sufficient buoyancy to float at rest and allow the thrust of the motor to raise the whole free of the water until conditions of planing were achieved.

Right up to this day the same basic

Fig 23 Gyro-controlled one-sixth scale model at a speed of 97.5 m.p.h.

planing surfaces have been used, although instead of skis the aft running surfaces have been surmounted by a float or sponson giving strength, buoyancy and a means of attachment to the hull.

We tried many "configurations" at this stage, but eventually a promising result was achieved by about April 1950, with the model shown in Fig. 20. This with others, one of which is depicted, was further tested at Haslar mainly to compare their ability to pass the "hump" speed with the known thrust available.

Students of planing forms will perhaps observe the absence of vee or dihedral in the planing surfaces. It would have been better in some respects, particularly from the viewpoint of reducing peak loadings, to have had some vee here, but up to the time we had to "freeze" the design in the interests of production, no really satisfactory behaviour—that is to say, freedom from porpoising—had been achieved in such a model, which all goes to show how near we are in all these cases to this trouble. To some extent it is a case of "easy come—relatively easy go!" An advantage on the score of resistance and therefore speed can be anticipated with the flat planing bottom.

About June 1950 it was decided that as the speeds we were considering in the full scale were so much higher than could be reproduced with any degree of accuracy in a tank model for various reasons, it would be desirable to reproduce the most promising of the forms so far developed up to one-sixth full size, when, besides hydrodynamic qualities the aerodynamic qualities could be expected to be faithfully reproduced at "scale" speed (Figs. 21 and 22). The main reason here is that the value of Reynolds Number for both air and water for the one-sixth scale model would lie reasonably close to and in the same turbulent range as for the full scale boat.

In order to reproduce the thrust of the full size job to scale, a rocket giving about 30 lb. of thrust for 20 seconds was aimed at.

This was developed, produced and calibrated for us by the Rocket Propulsion Division of R.A.E. Farnborough. We were very fortunate here, as without this fine piece of work we could not have driven the one-sixth scale model at the speed necessary to simulate full scale. This establishment even developed for us a thrust recorder against time scale, weighing only 6 oz.

However, in the early morning of 25 January 1951, a few of us were delighted and thrilled to see a 5-foot version of our successful little model literally leap into life from a model dock and run at high speed, controlled directionally by a gyro for 2,000 feet down the middle of an artificial "lake" 150 feet wide. Even more encouraging, the speed officially recorded was 97½ m.p.h., which, for a model, must surely be a record. All the time the model behaved perfectly. Those with a technical leaning can work out for themselves the equivalent speed at full scale, but it is around 240 m.p.h.! (Fig. 23).

This must not be taken to indicate we

Fig 24 *Crusader* under construction

anticipate reaching any such speed, because other factors come into the picture, such as the "scale" relationship of thrust to displacement, but it does give a strong indication that the behaviour of the finished craft should be as good as the model at this speed.

Later, John Cobb, Reid Railton and many interested experts came down and put the model through its paces in many high speed runs in varying conditions during 1951 and finally it was decided that the performance of the model was promising enough to justify construction of a full size boat.

Also on the strength of promising model tests Major Halford of the de Havilland Engine Company persuaded the Ministry of Supply to lend one of his Ghost jet engines of 5,000 lb. static thrust.

This then finished one phase of the design, namely production of a form likely to fulfil what is required of it.

Before concluding my remarks on the development of the form I must pay tribute to Mr. Brading, assistant chief draughtsman of Vosper, Ltd., who has worked with me on all these ventures, and who, with Mr. Hatch, also of our drawing office, has carried out the majority of the detail designing and hard work generally involved at this stage. Especial mention should be made of Mr. Taylor our model maker—his work has indeed been arduous as well as invaluable.

It can well be imagined that with free models travelling at speeds in the neighbourhood of 100 m.p.h. in open waters damage of an extensive nature was not infrequent. On occasions in bad weather I fear my impatience to test a particular feature was such that I would order the model to be released in enclosed waters knowing quite well that it was certain to collide with something before the rocket was spent. Mr. Taylor, however, was quite impassive and our little joke on these occasions, seeing the poor model in several small pieces, and knowing that restoration would take weeks rather than days, was to ask him whether he could be ready for another run that evening.

The structure consists of a planned mixture of birch plywood and high tensile aluminium alloy. The structural problem presented in the hull consists mainly in the acceptance without overstressing of the forces imparted by the water to the planing surfaces during the passage of the boat at high speed over a possibly uneven surface.

In the case of the sponsons large forces could be anticipated which would have to be transmitted via the cantilever arms to the hull, which in this region had of necessity to be as narrow as possible, while at the same time housing the turbine unit on its inside diameter. The sectional drawing will show how this problem has been tackled.

In the evaluation of the loading likely to be experienced by the planing surfaces we were given much helpful advice by Mr. Peter Crewe, chief hydrodynamicist to Saunders Roe, Ltd. Reid Railton took a strong line and very considerable in-

Fig 25 Internal view showing construction in fore-body

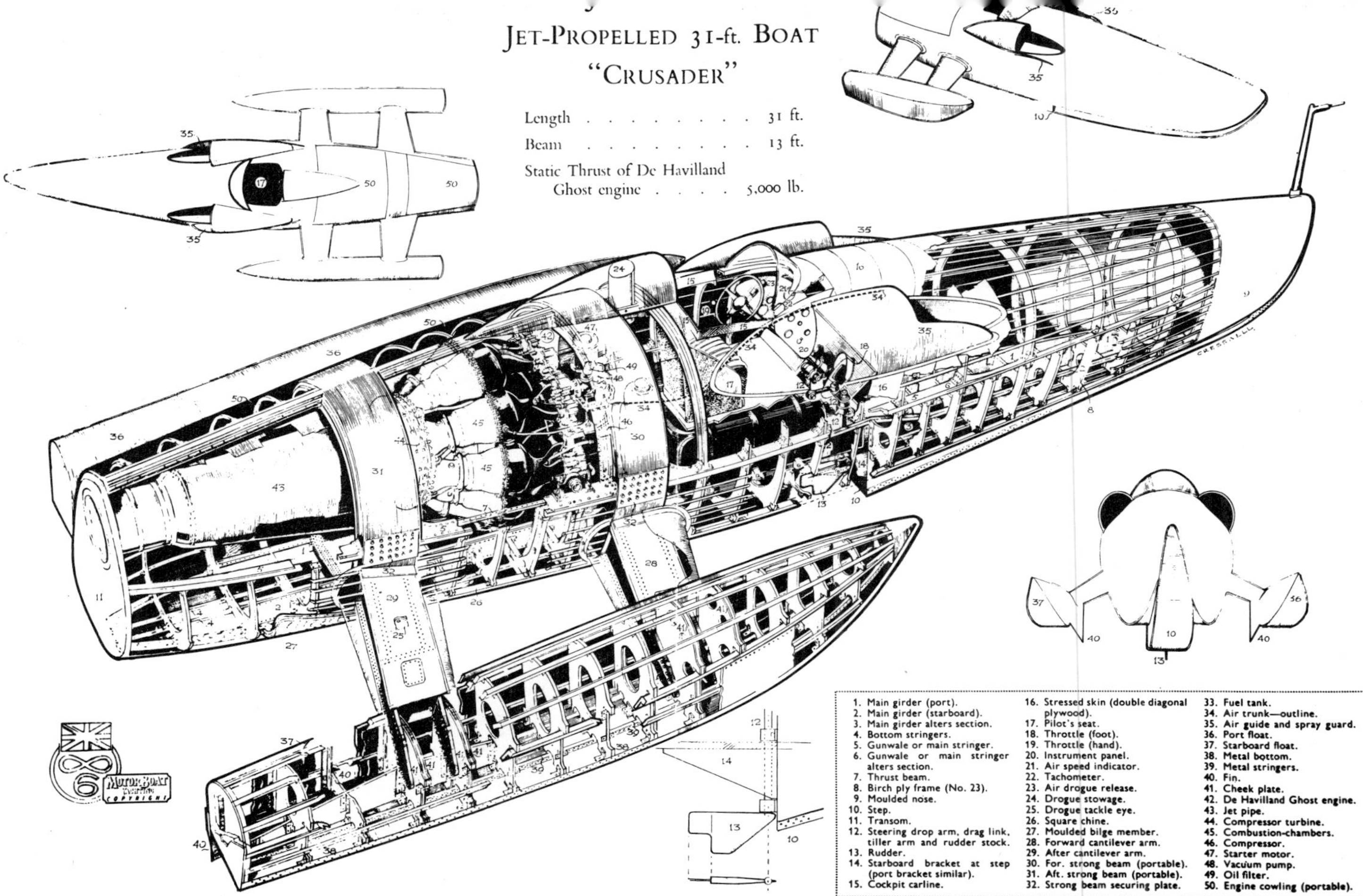

Fig 26 Cutaway drawing of the *Crusader*

terest in this point.

For the hull two skins of birch ply are fastened to a system of spruce longitudinal stringers, which in turn are "lugged" to birch ply framing sawn to shape. The lugs employed are in some cases of aluminium sheet and in others of plywood.

The birch ply, which has been supplied by Messrs. Wm. Mallinson and Co., is very suitable for this type of construction, being very strong for its weight. For the higher stressed regions and in particular on the planing surfaces and in the cantilever and hull ring members supporting the sponsons, aluminium alloy was used.

Very great assistance was rendered by Dr. Corlett and his company, the British Aluminium Co., who supplied the material and assisted in the design of the highly stressed aluminium members. Here DTB610B was employed giving an ultimate tensile strength of 32 tons/in^2. This is, of course, after solution and precipitation treatment. Proof strength at 1 per cent elongation is about 24 tons/in^2. In the case of the planing surfaces a sheet to specification NP5/6 was used having a proof strength at 1 per cent yield of about 14 tons/in.

It remains finally for us to observe the necessary humility in full appreciation of the fact that we have yet to achieve something for Mr. John Cobb in the way of a boat capable of record speed, to reward his very sporting spirit in fostering this venture.

The magnitude of the task is no light one and much development work remains yet to be done unless we are more lucky than is usually the case in these ventures.
(*Motor Boat and Yachting*, *Aug. 1952*)

Fig 27 The *Crusader* at speed on Loch Ness

Again we had the support of Dr. Gawn of A. E. W. Haslar and his staff; a large number of models were made and run in the experiment tank. There was, however, a limitation on the scale speed at which a model could be run so we used many free models propelled by rockets. This technique seemed good, so when we had produced a model at small scale, using the same technique as for *Bluebird II*, we built one of much larger scale and obtained a rocket from the rocket experimental establishment. This had near enough the thrust and weight needed to produce the thrust required to drive our 5 foot model at the speed that reproduced conditions likely to be achieved in the full scale boat. A.E.W. even went so far as to produce for us a gyroscopic directional control to keep the model straight in the torpedo trials basin at Horsea Island, which had the required length. We obtained a steady run timed at 97½ m.p.h. with this model (see Fig. 23) one early morning, and this seemed to all concerned, including John Cobb when he came down to see it run, good enough. The model gave the impression it would not be too difficult to reproduce at full scale, though the arms or cantilevers transmitting the planing forces from the sponsons would obviously involve some pretty severe bending moments and shock loading generally.

The basic construction was of plywood reinforced by aluminium where it appeared to be desirable. In fact, there was a good deal of aluminium throughout including a very late change to the use of it for the very small planing surfaces, which I think was possibly not the right answer. We had originally specified a good strong plywood panel which, for the same strength and weight, would have been many times the thickness. This in a uniformly loaded panel end held results in stiffness proportional to at least five times and probably more for the same weight. In other words the deflection at mid point would be much less. This was, of course, a most important consideration because somewhat against my inclination I had made the float bottoms and forward planing surfaces flat. This was done in at least an initial effort to eliminate the porpoising or steady state combination of pitching and heave so frequently met with in the really fast waterborne craft. It is analogous to, for instance, the flutter of tail or wings experienced at that time in the faster aircraft and almost equally lethal. As told above, we suffered in the gas turbine version of the *Bluebird* from this trouble.

The phenomenon was not particularly well understood at this stage of the art, although one or two learned papers had been written on the subject. In particular the head scientist of R.A.E. Farnborough had written a most erudite paper relating to the cause of this behaviour in Kaye Don's big record breaker *Miss England III* built and designed some years earlier by Thornycroft at their Hampton Launch yard on the Thames. So far as I am aware as a result of this paper the porpoising was cured—certainly the boat took the World's Water Speed Record of the day on Loch Lomond, but at a lower speed than the *Bluebirds* were doing. In fact I don't think the speed attained was much more than was achieved by our 5 foot rocket propelled model of the *Crusader* in the Horsea Lake, as described above. For all I know this is a record for a waterborne model to this day.

Our thinking on the subject of porpoising was to increase the angle of attack of the two aft sponsons relative to the forward planing surface, which seemed likely to prevent the fore part ever lifting to any substantial extent, and if by chance the waves encountered did cause a temporary lift of the forward part, the powerful moments set up by the aft planing surfaces would very soon force the fore end down and keep it there. Probably in fact this was too much the case.

It seemed to work well and certainly in the model there was at no time any sign of porpoising. But as will already be known from our efforts with *Bluebird II* with a "Goblin" installed, this is not, in itself, proof positive, though it is a guide. This time we had no porpoising with the full scale boat until the last few seconds of her life, when the forward planing surface had obviously become grossly distorted or actually fractured; but this is another story we shall touch on later.

To show how important is the aerodynamic lift effect in these craft the original *Bluebird* could not safely pitch upwards more than about 1½° without becoming dangerous, while the *Crusader* with her entirely different configuration could take up to about 4½° at her designed full speed of 250 m.p.h. These figures do not indicate much margin for either boat, especially the *Bluebird II*, but even one degree of pitch is quite a considerable movement in practice. There has, to my knowledge, been no official report published on what happened to Donald Campbell when he was killed in his later *Bluebird* at Coniston some 15 years later, but it seems highly probable that some disturbance caused this highly successful boat to take off and somersault in the air at about 300 m.p.h. The ciné photos shown on TV showed this clearly and all too realistically.

There must have been some unexplained or unusual disturbance because his designers had shown themselves many times to be highly competent and most painstaking; so they would never have allowed Donald to try for the 300 m.p.h. mark unless they considered it a reasonable performance. His boat was, incidentally, mainly of the original 3 point configuration where two planing surfaces are forward and one aft. He described the behaviour leading up to disaster as "tramping".

We have now reached the stage where we could build *Crusader* and test her on the water. As mentioned previously the cantilever arms fastening to the two aft sponsons representing the aft planing surfaces presented a formidable structural problem, in which we were considerably helped by Dr. Corlett, who was at that time a naval architect and consultant for British Aluminium Co. and is now of Burness Corlett & Partners at Basingstoke and very high in the naval architectural world.

We first ran the *Crusader*, or tried to, in the upper reaches of Portsmouth harbour near our works at Portchester, but only to check on controls and get an idea of whether she was likely to get through the hump condition and sit up and plane. The day was poor and the available water space limited; so after a number of efforts, during which it was clear that water was being splashed or sprayed into the turbine intakes, and we also found that getting up to plane was going to be obviously difficult, we stopped the trials until we reached Loch Ness, where there would be more room.

However, we had learned enough to take some measures to prevent the turbine intakes being flooded and to help her into the planing condition.

The next step was to tow the *Crusader* by road up to Loch Ness, which had been selected for the record attempt in view of its relatively great length. Here we had some more trials in late August 1952.

One morning at 4 a.m. it fell to me to get her up on to the planing condition because for some reason John Cobb was late and George Eyston did not want to miss the dawn calm. To this day I shall never forget the exhilaration of feeling the tremendous acceleration when, after a considerable bath from spray, she suddenly freed herself and gathered speed in a big way. This could, of course, have been expected, because in the Haslar tank, where we could only take the model up to perhaps the equivalent of 140 or 150 m.p.h., we had found the resistance/speed curve became a straight horizontal line after clearing the hump low down on the speed scale.

The inference was that if there were a constant margin of thrust over resistance the acceleration will be constant, and this was one of the important factors leading to the eventual unhappy end. The acceleration was so tremendous that, from analysis of the running in the last and fatal run, the *average* over one leg of the measured mile was 205 m.p.h. officially, and yet poor John had cut out the thrust *before* the half way mark. From this it is not difficult to estimate his speed to have been well over 250 m.p.h. when trouble started.

We are jumping ahead in our story of this record attempt. But I should like to say that I, as the responsible designer, knew of certain defects after a certain amount of running had been carried out, and beseeched John to let us have some time to put them right. I offered to take the boat back to Vospers at Portsmouth and effect the modification and strengthening necessary in our own plant at our own expense.

A high powered meeting was called at the Drumnadrochit hotel and I was a lone voice. Everyone, including Reid Railton who had seen for himself the trouble, was anxious to beat the record this year (1952) by a small but sufficient margin. Then afterwards we could use the winter to put things right and come up again the following year to put the record well beyond reach for some years.

Meantime, there was quite a lot of running done by John, who exceeded the record by the necessary margin quite easily on more than one occasion, but there were no official timekeepers present. Unfortunately it became obvious to me, and I fancy to most others closely connected with the record breaking attempt, that each time she ran fast the aluminium sheathing or panel forming the forward planing surface became distorted.

It will be recollected that I had purposely arranged the angle of attack of the two aft sponsons to be such that the forward surface was relatively heavily loaded in the interest of avoiding porpoising. Each time all hands got down to the work of providing extra stiffening inside the structure, but of course this was at best only a palliative. With hindsight I am certain we should have reverted to our relatively thick panel of birch ply which, for the same weight, would have possessed many times the stiffness to resist distortion. However, it was not to be, and if other vital factors had not entered into the last and fatal run John would have got away with it.

Eventually I was persuaded, most reluctantly, to agree to it running *provided* John would take the great care only to run in perfect conditions and not to exceed 190 m.p.h., which was quite enough to beat the existing world's record by the necessary margin. To this stipulation he agreed and I have a letter signed by him to this effect. The difficulty, as mentioned above, was to realise the terrific acceleration once the *Crusader* came over the hump speed.

In the meantime the weather had become bad and everyone was getting bored, including, of course, John and certainly most of the representatives of the press.

When, eventually, on the early morning of 29 September, it appeared the weather might be suitable for a few hours, we all stood by from 0400 hours. Exasperatingly, however, we found the weather nearly right, but not quite. Loch Ness is very long and the mountains or hills on either side are apt to cause local pockets of wind and rough patches of water, so we could not get the whole length of the course right. Thus it was until about 0900 hours when John sent us all to get some breakfast, in the middle of which it suddenly became flat calm and we were called back again in a hurry.

My station was about half way along the course on the overlooking high ground. With others I had to report to the control on the condition of the water opposite me. After a very short time I could see the *Crusader* out in position to start the run and it appeared to me the water on the measured mile was suitable. However, to my horror just after John had started a very fast run towards the measured distance a series of waves in a band of about 3 or 4 waves started to come out from the steep sides until they spread into the path of the *Crusader*. I watched this with growing alarm, and though one must suspect John could see what was happening he hit them with a crack like a pistol shot, then cut out momentarily and from that time onwards started to porpoise violently, never gaining control again, though still travelling very fast until after the end of the measured mile. Water, which one must assume had come into the hull through the forward planing surface, burst the whole boat open and she disintegrated. John was thrown forward out of the cockpit, but the violence was too much and he must have died instantly. From the fact that he averaged 205 m.p.h. over the whole measured mile and that he was first in trouble, as I have said, *before* the half way mark it is not difficult to realise he must have been going very fast indeed at this point. It would be difficult for him to keep an eye on the air speed indicator and he was undoubtedly going faster than he had meant to. The waves, possibly not more than 6 inches high, though probably more like 1 foot, would be too much for a boat such as this with an already suspect forward planing surface. So a great tragedy resulted and a marvellous man lost his life.

The film taken at this point shows clearly the band of bow waves which could possibly have been set up originally by the motor yacht going up to her station some 15 or 20 minutes before the *Crusader* ran over

Fig 28 Example of wave formations as a flotilla leaves Alexandria

the course. Certainly she went at full speed. The loch has steep rocky sides which will reflect waves with little of the attenuation which results where there is a beach. The typical wave pattern from such a ship can be seen in Fig. 28.

There was little comfort for any of us, and of course a number of different theories as to what happened floated around; but I found a sufficient number of witnesses whose statements all confirmed what I described above—not that this helps when such a marvellous man is lost. However, I have these statements to this day, one of which is reproduced here:

2 Holmfirth Flats,
Holmfirth Road,
Sea Point,
Capetown.
8th Oct. 1952.

Dear Comdr. Du Cane,

I had already left Aberdeen before your wire arrived, and it has taken some time to catch up with me—thus the delay in replying. Furthermore, as I am leaving London first thing tomorrow morning for Southampton, and home, this must unfortunately be a very brief and hurried report, or eye-witness account of the accident on the 29th September.

At the outset, I should like to point out that although I served for 6 years as a pilot in the R.A.F. during the war, and learnt quite a lot about thrust, drag, pressure waves, resistance, etc. as applied to air travel, I am somewhat ignorant of these same factors, when applied to a body travelling at high speed through water. This report, therefore, is purely *factual*, and I shall not attempt to advance any *theories* as to what happened during the fatal run, except to dispute one theory which I see has already been advanced, namely, that the three waves which the *Crusader* hit, were pressure waves set up by the *Crusader* itself.

Just before the start of Mr. Cobb's last run, I was standing in front of your car overlooking the measured mile, and at a spot, which I think you will agree, was rather nearer the South marker.

I was studying the surface of the Loch, which appeared from this position, to be absolutely ideal—glassy and no ripple. Just when the second Very-light was fired, however, I noticed a slow ripple emerge from the near bank, which I took to be a back-wash from the bow-wave set up earlier by the *Maureen* (I think that was the name of the boat which took the time-keepers out). When I noticed this, I expressed the hope that Mr. Cobb would keep well over towards the East side of the Loch.

The next moment he appeared, rather close to the West bank. It was a thrilling sight, and the *Crusader* appeared to be going splendidly. Just as it passed me, I concentrated all my attention on the floats, as I was particularly interested to see what happened when they hit these three slow, shallow swells.

There were three distinct bumps, and although the *Crusader's* stern remained in the water, the front bounced rather viciously. I don't think the *Crusader* started breaking up at this stage, and as far as I am concerned, the rest is a matter of conjecture. I rather think that John Cobb reacted to these vicious bumps in the same way as the average motorist would react to a speed wobble, when driving a car—he throttled back. The *Crusader* then appeared to nose-dive into the Loch before breaking up.

I trust that this report is satisfactory, and in conclusion I should like to say how sorry I am that such a gallant British venture had to end so tragically.

Yours sincerely,
(Sd.) D. Tyndall.

The late Donald Campbell told me a year or two before he was killed that our experiences with John's boat were of considerable help to him when he came to take up the torch with wonderful courage and persistence.

It is now more than 4 years since Donald was killed at Lake Coniston, after a brilliant career on water and land and after raising the world's record many times before the final disastrous run when he was aiming at 300 m.p.h. I have nothing but admiration for Donald, though I was not able to work satisfactorily with him during the short period he treated me as his expert adviser. This was, to some extent, due to a clash of temperaments and because I did not really believe in restoring the old Rolls Royce piston engine to his boat. However, I wrote out a series of trial runs for him to make progressively at Coniston when he started trials, which is the way I always approach this type of development into the unknown. If one is the designer or responsible for the technical aspect of the attempt, it is frankly the only way to approach the job. There is hardly any difference between this type of development and the way the aircraft companies approach the problem of testing and eliminating the bugs in an advanced, sophisticated and fast type of aircraft, be it fighter, low level bomber, or high performance passenger liner. We now have hovercraft to bedevil our technicians with prototype troubles. As a matter of fact they are all pretty good these days, though even these beautiful machines nearly always suffer from a series of troubles, sometimes serious, in the first year or two of their commercial operation. In one important way we are much more circumscribed by lack of finance and facilities.

In the case of an aircraft firm—certainly one with the modern and highly sophisticated approach to the design problem—one or more chief test pilots will be employed to carry out the trials ordered by the flight control officer in conjunction with the chief designer, in a written and detailed fashion. Being employees of the company they will not hesitate to carry out the programme to the best of their ability and they are trained to possess the necessary observational faculties, and to have the ability to write up an understandable report.

This is not to say that a test pilot will not make suggestions based on his experience in the air, but the chief designer or flight test engineer, as they are usually called in the U.S.A., will have the last word. Incidentally, it is extraordinarily satisfactory for the test pilot to have undergone an advanced engineering course, because he is then even more likely to be able to explain and certainly to write, a good report on what he has experienced. In the U.S.A. a great many of the really star turn test pilots, whether they are employed by a Government agency or by one of the big corporations, will have a degree from one of the better known universities specialising in engineering, such as M.I.T. (Massachusetts Institute of Technology) or Caltech (California Institute of Technology on the West Coast). There are, of course, a number of others where a good engineering degree can be obtained at Government expense.

Many will undoubtedly say that a degree is not a necessity for a test pilot and this is obviously true, but at least it helps a man to write a logical and clear report following his experiences in the air, where he may be alone. The development of the X15, by both firm's pilots and the military pilots subsequently from Edwards Field in the Mojave desert, is a case in point where the test pilots had engineering degrees. And what an achievement!

I never really got to know Donald Campbell well. After we parted by mutual agreement he took on Dr. Corlett, who looked after him for a time and tried to obtain some more speed by making use of the principles of the surface propeller, which had been developed with great success on the West Coast of the U.S.A. in a recent record breaker *Slo-Mo-Shun*, designed by Ted Jones. I believe he had difficulties also.

I do not doubt that given time this would have been successful in pushing up the speed of the *Bluebird*, but something happened to the shaft or shaft log and to avoid actual sinking Donald beached her on the shores of Lake Coniston, where she lay—more or less a hulk—for some years. Donald, with his outstanding courage and persistence then commissioned a jet boat designed, I believe, in the first instance by Dr. Corlett and taken further by W. Crago who was then in charge of the Saunders Roe model tank. I am not aware of the sequence of events, but certainly a very large part was played by the Norris Brothers, who ran an engineering firm in Sussex and seemed to have the necessary background of experience for designing and building record breaking boats and cars.

After many disappointments and much financial worry Donald eventually put the world's record well beyond the reach of any piston engined boat, and continued over a space of many years to improve on this speed until his final tragic run in January 1967. Then, after what sounded on the radio to be a pretty dicey run, he took off and turned on his back—still apparently talking to his control boys, which must have taken some quite exceptional nerve considering he must have realised he had little hope of surviving. This may or may not be correct but it was reported at the time, though not in a book recently published.

The movie photos of this stage in the run were quite horrifying, and

Donald was never found after the crash and sinking of the boat. Only some bits of the boat were salvaged. A very courageous and persistent man was lost who, I fear, despite the publicity surrounding most of his attempts on land and sea, tended not to be properly appreciated.

To go in for this sort of thing there has to be something out of the ordinary in your make up which the majority of more cautious beings do not have. To conclude my remarks on the redoubtable Donald, who was undoubtedly as good as, if not better than his father, one cannot omit to mention the extraordinarily valuable services rendered throughout by Leo Villa, their dedicatedly loyal, and most capable advisory engineer, who eventually became an essential and highly respected part of the family and who has since published a book. In it he recounts an amusing experience when he tried to get Malcolm out of bed to run the *Bluebird* at Coniston, but Malcolm was reluctant saying he was not feeling well. Leo Villa then said I was wanting to drive her, and Malcolm was out of bed like a shot!

Fig 29 The *Crusader* at speed

5

Some Yachting

My earliest experiences in yachting were crewing for the famous Ponsonby brothers, who were distinguished in the Royal Ocean Racing Club in later years. Both their father and mine lived at Dinard opposite St. Malo, and I was taken by them in their 5 ton cutter out to sea—at times as far as the Minquiers rocks; but frequently through those rock infested waters to St. Briac or further afield. One did not learn much celestial navigation in these waters, but a lot of pilotage and all round seamanship.

In 1924 I was loaned the Brixham trawler *Salem* while I was doing courses at Greenwich. She was the first yacht I had actually commanded and navigated across the channel, and was owned by Bill Leeds, the American tin-plate king, an all round sportsman and one of the most generous and fun loving people I ever knew.

My crew was not strong in navigation but plenty of fun. Being on the long summer leave from the R.N. College Greenwich I had plenty of time to spare so I got the good ship under way from Woodnutt's yard at Bembridge with the help of one Brixham hand and a local boy. The big idea was to go to Deauville with Tony Bellville, the "Rat" better known as the Baroness de Forest, Rupert Bellville, David Drummond and "Babs" as chaperone. I anchored off Southsea beach and shipped all the baggage and passengers aboard in the dinghy or one hired from a longshoreman. We were ready to sail by about 6.30 p.m., and I had to tell the Brixham hand we were returning to Bembridge as he refused to go "foreign" having no faith in my ability to navigate.

I was the only navigator on board, which put a strain on me, my not wanting to have to stay on deck all night. I asked if anyone knew the rule of the road at sea with the idea that we could put one of them on watch so that I could have some sleep. David Drummond, who, incidentally, had been in the Royal Flying Corps during the war and had two artificial legs as a result of trying to slow roll a Camel too near the ground, offered to keep a watch as he knew the rule of the road from flying. This was not entirely reassuring to me, but I went below leaving David and Rupert on watch. Rupert was sent as a look-out up forward.

After an indefinite period of sleep I was woken up by shouting on deck. I heard David say to Rupert: "get forward to your post of duty Rupert. I tell you there is no danger—red light means she is going away from us". At this I had, perforce, to come on deck and put her up into the wind to avoid a collision with a large ship in the usual steamer lane, although she should have given way to us as we were showing the lights of a ship under sail. But the motor was running as well. Perhaps they knew this!

We eventually reached the Havre Lt. vessel when the engine elected to pack up and nothing I could do would get it going again. It was, therefore, necessary to sail the rest of the way, including into Deauville harbour itself. I have now done this some 50 times, but back in 1924 this was my first approach to an entrance which is by no means a straightforward one. Luckily the wind served us well enough. Eventually, it turned out, too well, for when I gave orders to lower mainsail and mizzen the crew had jammed the halyards somehow and the mainsail was on her far too long, so that on passing through the first gate to the outer yacht basin we had obviously too much way on. The Capitaine du Port seeing the speed at which we were travelling in the direction of the inner basin, the access to which was closed, shouted while running with his men towards the second gate to the inner basin "En arrière—en arrière avec les machines". I replied "nous n'avons pas des sacrés machines qui marchent, ouvrez donc les portes et nous essayons ancrer". They just got the gate open in time, and it still took us the length of the coaling basin in which to anchor and stop. To pay us out he gave us the berth under the coal heaps.

After several days in Deauville most of the party returned home leaving me to take the yacht round to St. Malo with only the old man, the boy and a naval friend of mine, Ikey Mandeville, who was in the advanced engineering course with me at Greenwich. We did all right until we found ourselves in a bank of fog while about half way down the Chenal des Jardiniéres, just outside St. Malo. There was little one could do except drop an anchor as quickly as possible, this being well before the days of radar. When the fog cleared we found we had been carried out of the channel and between two dangerous rocks in which this part of the world abounds.

On arrival we anchored first of all in the Anse de Dinard, but had to come down when a gale blew and take her into the shelter of an inner basin in St. Malo. Here she lay three weeks before I could raise a scratch crew of naval officers including, amongst others, Georgie Spencer who had taken in a big way to the vodka served at the Cafe Caucasian, where we sat out most nights. Georgie was also doing courses at Greenwich and I had been shipmates with him in the *Hood*. My recollection is that after sailing it was some 36 hours before we were off the Isle of Wight, when the old salt suggested that as it was dark and low water we should heave to and await the dawn and tide before entering Bembridge. To this advice I was entirely agreeable; so it was not till about 1 p.m. the next day that we entered Bembridge.

This was my first and only experience of a Brixham trawler. They were fine ships for the job they had to do but the gear was dreadfully heavy and we lacked the usual steam donkey engine which the real sailing boats had for handling sails, anchor and so on; for some reason it had been taken out when she was converted to a yacht and had an engine installed.

The next yacht of which I had some experience, including part ownership with Tony Bellville, was the *Jolly Roger*, a simple conversion of an ex German patrol boat of World War I, about 70 foot long fitted with petrol driven M.W.M. engines. She was a good sea boat and fairly fast and we took her for her first cruise up to Paris, where we met up with Bill Leeds who gave us a marvellous time. There was an "exposition" on at this time and much fun was had by all including a free fight in a dart stall and a somewhat drunken challenge from Rupert, during dinner aboard a barge moored alongside the left bank, to swim across to the other side for £5. I regret to say I was beaten and became much hampered by my trousers, which I took off about half way across. We ended up in a very smart American club barge being hauled out of the water by waiters who, seeing that I had no trousers on, quickly put one of their aprons round me and led us off through the restaurant looking liked drowned rats.

After we had changed and warmed ourselves internally, if not externally, our host Bill Leeds made the following announcement as to our next move. He said "Say boys, I regret to tell you the first item on our programme is cancelled. The hermaphrodite is indisposed—we will go to the Papillon as a start". And so the night went on we visited most of the "Maisons bien tenues . . .". The Grotto and perhaps best of all of its kind Cent Vingt Deux, Rue de Provence. Here one could choose one's room according to taste. There was one decorated with a railway atmosphere as a sleeping car, though I do not believe the reciprocating motion, slight as it might be, was reproduced nor was the effect of passing over the points at a junction. An engineer could have helped them here, but they never asked me and strangely enough I was usually better occupied. The other room which I well remember was designed to represent a farm yard haystack with real hay. One day coming out of the Ritz Bar in the Rue Cambon, Napier Alington, one of the most charming and sophisticated of people, saw a car load of hay driven by and asked the driver if this was for Cent Vingt Deux. Needless to say the answer was in the affirmative with many jokes and mirth generally.

Bill Leeds had a dark girl with him who was introduced to us as being of very good Virginian stock and a perfect lady, but this did not prevent us ending four in one big bed at the Hotel du Terminus, as by this time I had found a young lady called Simone. Needless to say the old *Jolly Roger* was made fast at the Port de Plaisance between the Pont de la Concorde and the Pont Alexandre III, but I had to throw myself on Bill's mercy for a bed that night because I found my bunk already occupied by another member of the crew and one of the girls from the Papillon, who seemed to like being whipped. The whole party came down to L'Ecu d'or up the River Marne the next day when Bill produced the marvellous negro band.

I have no very accurate recollection, nor do I suppose has Bill Leeds, but I am sure he spent more than £1,000 that night in entertaining all of us. Needless to say the disgraceful party described in Paris took place long before any of us were married.

Gradually most of the party filtered back to England leaving Ikey Mandeville, Geoffrey Branson and I to bring the boat home with Abe Bailey, the superb engineer we recruited from London. He was a cockney of the best sort and was a wizard on the engines. During the worst gales he could be heard whistling cheerfully in the engine room. Geoffrey Branson was doing gunnery courses at Greenwich and Ikey was in my "dagger" engineering course, so we were all on leave at once. Our intended destination was St. Malo to see my father and stepmother, but in heading up for Cap Barfleur a gale from the south west blew up so I changed course for the Isle of Wight where we arrived the next morning fortified by brandy through a night that was not altogether pleasant as the wind must have been blowing at least Force 7. We entered the harbour in Bembridge and were well received by the Bellville family who had got back about a week earlier. This was the last I saw of the *Jolly Roger* because I was appointed abroad soon afterwards in the Royal Navy to HMS *Malaya* and then to HMS *Emerald* in the Far East.

One of the more amusing characters we took with us up to Paris was Vere Ker Seymour who was, I suppose, what one would call a roué, but a pleasant and amusing one, though much older than us. He wrote a book shortly afterwards in which some of our exploits in the *Jolly Roger* were described. He also recounted a wrestling match organized about this time in Paris between a German and French champion. This match was heavily sponsored by the betting members of the Travellers' Club and was of the nature of a Free for All. Posted up around the ring on large placards were the following two regulations:—

"C'est defendu:

(1) Le prise aux yeux

(2) Le torsion des parties sexuelles."

I did not see this match but was told afterwards that the Frenchman was disqualified for infringing the second regulation.

While I was abroad serving in the Mediterranean station and then China and East Indies without returning home, Tony Bellville sold the *Jolly Roger* and that was the last I heard of her. Dear old Tony had some most convincing story to explain why my share in the *Jolly Roger* was worth nothing—and nothing did I get for her. However, Tony was a kind and amusing friend and I do not doubt my share of the money went in some amusing and worthwhile enterprise.

On returning from 3 years abroad in HMS *Malaya* and then HMS *Emerald* I took a pilot's licence at Brooklands and hunted with the Fernie on horses provided by Frank Bellville, Tony and Rupert's father. This proved to be a fascinating way of passing the winter months and I have managed to hunt in Leicestershire or Northampton for upwards of 40 years since.

The next yachting I can remember was after I had joined Vosper in 1931 and had sold Tom Rose Richard's cutter *Iolaire* to Jack Alston, who lived at Hillhead. Jack was not very experienced but the *Iolaire* had made a name for herself in ocean racing in the hands of Tom Rose Richards. So we decided to have a go at some of the more amusing ocean races after we had done a bit of cruising in her, which included a visit to Bobby Somerset's home Ince Castle on the river Lynher—a tributary of the Tamar running into Plymouth harbour. Strangely enough Victory, my wife, had been brought up opposite Ince at Antony, which was the estate of her father and then her brother Sir John Carew Pole, who succeeded him. A beautiful part of the world, probably only excelled by the south west coast of Ireland where we now live at Glandore, Co. Cork. Here, overlooking the harbour, we can see the big tankers making their way to Bantry, where there is sufficient water for them to unload at Whiddy Island. We see too the occasional liner passing between Galley Head and Mizzen Head on their way to New York.

The first ocean race in which we took part was the Channel race of 1936, when we had to give up, Charles Courtney injuring his spine during a gale which sprung up in the night after we had rounded the Royal Sovereign Light Vessel and were making for Le Havre. We found ourselves pretty unseaworthy when the gale got up and it was only with the greatest difficulty we managed to shorten sail, when I found one of the two fairly useless crew

Fig 30 *Seamark* Glandore Co. Cork

Fig 31 View of Glandore Harbour from *Seamark*

on his knees in the forecastle praying instead of helping us. We made Dieppe which was directly to leeward.

It was not until the following month when the Dinard race was on that we qualified for the Royal Ocean Racing Club. On the whole I think we deserved it. In those days we were allowed to go between the Channel Islands and the French mainland, and what with the Alderney race and the multitude of nasty looking rocks the navigation, or pilotage, was far from being just a "piece of cake", especially in a fog or low visibility. We took with us Victory (my wife) and Audrey Bellville who had a pretty rotten time because, owing to the weather, we kept them in their cabin most of the time where the fumes from the adjacent petrol tank made them feel rotten. As it was a race we had to drive the boat as hard as possible, and the ship was well heeled and throwing herself about in a near gale. Having no life lines rigged we were genuinely afraid the girls would be thrown overboard. Possibly we were unkind but it was a serious risk we did not want to take. We finished the race not very high up the list, but sufficiently well to qualify for the Royal Ocean Racing Club, with its useful club house and attractive crest. I had found *Iolaire* rather a handful to steer in a following wind up the entrance channel to St. Malo, which is where some years before we had nearly come to grief in a fog bank with the Brixham trawler *Salem*.

The seagoing part of the race was to some extent disturbed from complete harmony by the extraordinary state of mind of a naval captain (retired). He took a hate to me, who did a good part of the work, and while

I was out on the end of the bowsprit at the start he released the preventer backstay and accused me of doing this, which was of course totally impossible. When I took over the watch from him at midnight he seemed to lack confidence in his navigation because his last words to me before going below were "If you see breakers ahead put her about". I really believe this friend had temporarily lost his nerve, for the visibility was quite good and whatever else can be said about the Channel Islands area the lights and navigational aids in peace time are excellent. I had no trouble, though being a race one could not use an engine, without disqualification, to get away from a dangerous position.

Since the war this race has been run with a prohibition on the competitors entering the area between the Islands and mainland. In other words, Alderney, Guernsey, Jersey and the Minquiers rocks have to be left to port.

Our yachting activities after this race were curbed until the war owing to the greatly increased necessity for me to attend to a rapidly expanding business at Vosper. My wife, however, managed to go with Tom Rose Richards in the *Lexia* on a Belle Isle race and I managed one cruise with Tom to Plymouth across to Brest and back to Chichester, his home port.

No account of my yachting activities would be complete without mention of Loel Guinness who had been one of our most faithful customers from the time I joined the firm in 1931 to the present day.

In the early thirties Loel had the splendid little steam yacht *Carmela*. She was a coal burner with double expansion reciprocating engines. To give an idea of Loel's generosity, he lent me the yacht for a week end, and my wife and I, Tony and Audrey Bellville, and Captain Hugh Woodward got up steam and sailed her from Bembridge to Poole where the pre war motor boat racing was taking place under the aegis of the Royal Motor Yacht Club. We left Bembridge before dawn and poor Victory sat on the flukes of an anchor in the dinghy which was very painful and led to an operation on her coxis or tail bone eventually. We had good visibility and managed to reach Poole soon after breakfast carrying the tide with us.

All went well for the rest of the week end and we ended up in Portsmouth harbour at the Vosper Hardway moorings on Monday. This was only after I had found it necessary to persuade the man at the helm that the Swashway was a perfectly good and safe short cut into the Portsmouth main channel.

After a time, about 1936, Loel found he wanted a larger and more seaworthy yacht with which he could cross the Atlantic if necessary. I was commissioned to look around and came across a pretty modern and attractive (for those days) proposition. This was the 296 ton diesel yacht *Atlantis* which I had sold quite recently to Lord Michelham, who did not find much use for her. It fell upon me to take the *Atlantis* out of the Camber for the first time after we had completed a survey and carried out a certain amount of work on her. Normally this would have been a fairly straightforward undertaking as I was in good practice at that time with the handling of the much more difficult single screw steam yachts into our Camber yard and onto the Hardway moorings.

However on this occasion, having a perfectly clear run out of the Camber from under the crane, I rang down slow ahead both. One started and the other did not with the result she took a sheer right across the Camber and I had to stop both. These engines, I should explain, were of the direct coupled type. When an engine started or stopped, so did the propeller shaft. This is all right as long as the engines do what is asked of them, but most yachts nowadays have a clutch and reverse gear between engine and propeller shaft. We eventually got her running on both and went out for a trial cruise. This was pretty good except that the hull of this yacht was out of trim—down by the head—and had only a rather miserable centre rudder placed between the propellers. She was, therefore, directionally unstable and not at all pleasant to handle, especially in a following sea. However, eventually Loel took her across the Atlantic and back so she cannot have been too bad, though in a gale most of the ship's company, except Loel himself, were pretty miserable and somewhat frightened; this included his wife, the former Lady Isobel Manners, whom he had just married. Isobel complained that this was not what she called a honeymoon and who can blame her!

Then came the war. Loel joined up with the Air Force, and the *Atlantis* was sunk on a mine while serving in the Royal Navy.

6

Vosper and Early War Days

After this diversion to discuss the two world record breakers with which I have been associated and also some yachting experiences, we must return to the tale of Vosper in the last year before the war, that is 1938–9.

By this time it was likely we should be involved in another war with Germany, and from our own point of view things in the yard were good, as we had received a further order for MTBs from the British Admiralty based on the experiences gained from the running in service of MTB102, our private venture craft with Isotta engines. In these craft, most of which were to distinguish themselves in action at a later date, modifications were incorporated based on "102". They were also slightly longer and possibly not

Fig 32 The 40-ft Royal Barge passing the Royal Yacht *Victoria and Albert* aboard which she was hoisted. She was designed and built by Vospers in 1938

Fig 33 The same 40-ft Royal Barge was later transferred to the Royal Yacht *Britannia*

quite so fast, but undoubtedly represented an all round improvement from the operational point of view. In these craft the bridge or control position was located outside and was provided with some armour.

We had also to undertake a rough weather trial and by virtue of modifications to the hull form a fairly considerable improvement resulted.

At about this time we were much pleased and honoured to receive from the Admiralty an order to design and build a new 40 foot Royal Barge for hoisting into the Royal Yacht *Victoria & Albert*—she was later transferred to the *Britannia*. This we accomplished in little more than 4 months and seemed to give much satisfaction both from the point of view of looks and performance.

The Royal Barge fulfilled her job for about 25 years and I must admit we were all very disappointed to lose by a close shave the job of designing and building the new one in 1965–6. But it may be of interest to see an artist's picture of the craft we put forward. (Fig. 34)

An interesting small fast boat designed and built for Lord Louis Mountbatten was produced about the same time as our Royal Barge (1938) to be hoisted on board HMS *Kelly* for his personal use. The feature which at that time represented an advance technique was a sliding top or shelter made up in perspex and aluminium, which can be seen in Fig. 35.

The draughtsman with whom I worked on this job was one John Rix who had just joined us from an apprenticeship at Thornycroft. He is now the chairman and managing director of the operating company Vosper Thornycroft Ltd. This arose because when I sold out my control of the company it eventually came into the hands of David Brown of much industrial fame. When David acquired control I considered it time to hand over the responsibility of managing director to John Rix with David Brown's approval, of course. It was shortly after this that Thornycroft were acquired and what amounted to a new regime commenced. John Rix is most people's

Fig 34 Design of a proposed new Royal Barge. From the painting by David Cobb

Fig 35 20-ft Speedboat for Lord Louis Mountbatten

idea of the managing director par excellence and I think we are lucky he has not been taken away from us as he is God's gift to some organisation like M.S.L. (Management Selection Ltd.).

This particular boat for Lord Mountbatten would not appear to be particularly dated even today, 33 years later. The same applies to the Royal Barge built well before the war. This perhaps shows one of two things.

Either we were most satisfactorily advanced at that date after only about 7 years of experience in design, or we slowed down as the years passed and made less, but still some, advance in the matter of appearance. Living, as we do in our yacht *Dimarcha* at a berth at the splendid Hamble Marina, and seeing fast motor cruisers or jolly boat types passing us up or down the river at intervals of 5 minutes or less, I cannot honestly find it possible to admire the majority of them. This is in 1971. Strangely enough the ubiquitous Chris Craft, which some years back were somewhat mundane types, have in the last 2 years evolved into some of the more attractive designs. This may be because they have retained the services of one or more of the high class U.S. designers, or naval architects, and most probably a first class stylist as well—they are, I understand, using the services of Jim Wynne of offshore power boat fame.

As soon as the war with Germany broke out for the second time the Royal Navy decided that a large number of the MTBs, on which we had been working for about 3 to 4 years, were required for naval service. But as we relied on the Italian Isotta Fraschini marine engines, giving about 1,150 BHP max., things looked awkward as Mussolini was likely to sympathize with the Germans. If he could have kept a calm outlook Italy was then in the most wonderful position to improve her economic position by supplying much needed armaments to both sides. I was sent out by the Admiralty in the first few days of the war accompanied by my co-director Lord Hardwicke, to place a very big order for the 1,150 HP engines, as well as for the 1,500 HP version which was, at that time, well on towards the point where it could be used in naval service. Our friends at Isotta in Milan seemed genuinely pleased to see us and were happy to take our large order and give the early delivery we required.

However, little actually reached England before the Italians declared war on us after the fall of France in 1940. The only significant thing I can remember about this meeting was the extraordinary experience of hearing the Isotta managing director ring Bosch in Germany, our enemy, about some generators which had not been delivered for us on time, and giving them a good ticking off.

Much progress in the delivery of boats already ordered was achieved during the war before Hitler attacked France, and a number of Isotta engined boats were delivered to the Royal Navy, who took good advantage of the phoney war to train their personnel for the real thing. My impression is, however, that with the exception of a few of the more far seeing of the younger officers, the Royal Navy as a whole did not think much use could come out of these little fast craft. But they changed their minds radically when, in about 1942, the little warships were the only ones which could be operated by the various Commanders in Chief of the chief naval ports on the South Coast. Those based on Dover fought battles almost every night.

The enemy did not use their big ships in this area, but they were dependent on keeping the convoys of supply ships running along the coast of the Low Countries, and France too, in order to help maintain the occupying forces, which could not rely on such railways as were left to carry sufficient supplies of all sorts. These convoys were vulnerable to the MTBs though well escorted by warships up to small destroyer size and capable of running with the merchant ships inside the many sand banks off the coast of the Low Countries.

In most weathers, and particularly when the nights were long, our coastal forces had speed enough to get across the North Sea or the Channel under cover of darkness much of the way, wait to make an attack, and return at least some of the way before it became light. As our new boats, fitted with Packard engines of the 1940–1 class, were unreliable at first, it was suggested to me it would be a good idea if I accompanied our MTBs on some of these missions to see at first hand what went wrong. It was about June 1942.

This was probably a help to the operational ships' companies and certainly to us and I like to think that matters improved progressively from that

Fig 36 73-ft wartime Motor Torpedo Boat

time onwards. Of course there were faults on both sides and perhaps in the Admiralty, but none the less no one could say we were free from various tiresome causes of unreliability and the installations were initially too complicated. They were much simplified later.

I should say that the few original Isotta engined MTBs, which had been ready for service at the outbreak of war or soon after, did not suffer to any extent from this unreliability, and acquitted themselves extremely well in these early hard fought actions. A good example was MTB31 commanded by Lieut. (now Commodore) Jermain, D.S.C. Unfortunately there were too few of them. Later on the Packard marine engines supplied by the U.S. under lease-lend became very reliable and all parties came to know how to instal and run them; but we went through an unsatisfactory and frustrating time for some months. Here again the faults did not lay fair and square on us, the shipbuilder responsible for the installation, although it became easy to blame us, for we held the overall responsibility. Eventually all was well and many highly satisfactory engine installations were carried out in our own and other boats giving excellent performance. It was, furthermore, a great advantage when the Packard was uprated to 1,500 HP, which gave us the ability to maintain or exceed the previous speed of about 40 knots while carrying the ever increasing armament load called for by the naval staff. This is shown by accompanying table:—

COASTAL FORCE DESIGN

70-FT VOSPER M.T.B.S (SHORT TYPE)—COMPARISON OF MILITARY LOADS

Class of M.T.B.s	1938	1939	1940	1942	1943	1944
Admiralty numbers	20–23	31–40	73–98	347–362	380–395	523–537
Petrol, in lbs	11,100	14,120	18,818	19,055	18,870	18,870
Lubricating oil, in lbs	435	610	1,090	796	1,525	1,100
Fresh water, in lbs	500	930	450	500	450	450
Crew and effects, in lbs	2,160	2,400	1,800	1,800	2,160	2,880
Miscellaneous items of equipment which constitute additions, in lbs	360	713	703	809	1,384	1,572
Armament, ammunition and armour protection, in lbs	12,860	14,805	15,147	16,294	18,362	19,000
W/T and radar or A/S, in lbs	540	927	836	1,064	1,800	3,632
Total weight, in lbs	27,955	34,505	38,844	40,318	44,551	47,504
Per cent increase in military or useful load compared with 1938 M.T.B.s	—	23·2%	38·5%	44·0%	59·2%	69·5%
Total displacement on trials, in tons	35·79	39·70	46·90	44·73	44·39	48·80
Per cent of total displacement made up by military load	34·9%	38·8%	36·9%	40·3%	44·7%	43·4%

On one occasion when the war looked like coming to an end within foreseeable time we asked the Admiralty for permission to run one of these boats as light as possible and using some 15 per cent more power out of the machinery, just to see what speed we could achieve. The result was 49 knots average over the Stokes Bay measured mile, which was highly satisfactory for those days and never equalled until the gas turbine engined boats came on the scene in the 1950s.

The war was not really a satisfactory period for me, and I suspect, for most of us. We were entirely under the thumb of the Admiralty and it became practically impossible to do anything in the way of obviously desirable modifications and improvements without their approval, which was quite often not granted. Compared to the pre-war years when development was pretty fast and we were left much to ourselves this was frustrating, though one can understand why control was imposed on us in the interest of standardisation.

To give one example, it was not long before we came to the conclusion that the hull form could be much improved in certain respects. Before the war we had the advantage of making use of the Haslar Tank and Dr. Gawn's advice in testing a large series of forms in an endeavour to obtain the best compromise between speed and seaworthiness. The final result seemed pretty good and the form was, in fact, standardised throughout the war on our boats. However, as the operational use of these little warships settled down to a pattern, it became quite obvious that most operations would be carried out at night, and that to attack one of these north-south convoys running from German ports to the Low Countries and France it was essential to patrol at fairly slow speed, say 15 to 20 knots, using the radar to pick up the convoy and to get into the right position to attack with torpedoes. In this part of the operation the utmost silence was essential, the enemy having many listening posts ashore, and it was necessary for us to approach pretty close to be in a position to intercept and attack.

Unfortunately at these low speeds the chine forward became immersed and threw a good deal of spray over the bridge and forward guns, which was most undesirable. This was not the case at high speed, but except during or after an attack high speed was seldom used. The fault could have been cured quite easily by raising the run of the chine forward, but until the end of the war we were not permitted to do anything about it in the interests of standardisation of cradles for slipping and other details. In fact MTB1601 was the first in which we were allowed to use a modified hull form.

A good deal of effort was put into silencing the Packard engines, and certainly the operational boys achieved many successful attacks. However, due to the delay in production caused by Italy entering the war and the complete stoppage of Isotta engine supplies, there was a period of nearly 18 months during which little could be done except to await the Packard supplies from the U.S.A., though there were a few hulls fitted with another engine from U.S.A.—the Hall Scott of considerably less power. This enabled a certain amount of operational work to be carried out in the really narrow seas, but under severely handicapped conditions and mostly from Dover owing to the limitation of speed. The boats with a maximum of probably not much more than 30 knots could not operate from bases like Felixstowe, Lowestoft or Yarmouth owing to the time taken to get into the attacking area and back before being caught out by enemy aircraft in daylight.

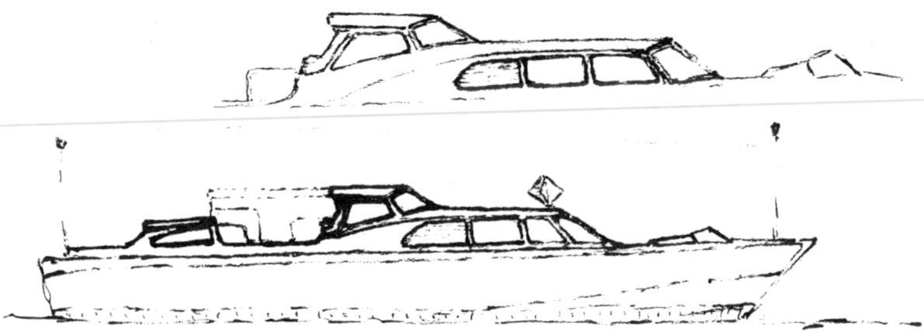

7

Fleet Air Arm

It was during this period of awaiting the engine supplies from the U.S.A. that I joined the Fleet Air Arm as an air engineer and pilot having been persuaded by a friend that there was a considerable requirement for officers with both of these qualifications.

By far the most difficult part of aviation at that time was flying across country, particularly, in bad weather and low visibility. Nowadays, of course, this is in many ways easier because there are so many radio aids and ground controls, etc. Also the ability to fly through clouds or low visibility of any kind is greatly assisted by the various blind flying instruments and auto pilots, provided one has a reliable radio. True, it takes a good deal of skill to land a modern jet airliner of the Boeing 707 or similar vintage, but the equivalent of the aircraft we learnt on in 1928 almost land themselves nowadays. The tricycle undercarriage has made a lot of difference to ease of landing and control generally on the ground.

Having volunteered for service in the Fleet Air Arm I was first appointed to Eastleigh for flying instruction and conversion to Fleet Air Arm types and to be brought up to date, as it was several years since I had flown. I think the course lasted 3 weeks and very good value it was, although I found myself pretty unskilled at first. Amongst other things. Service aircraft had changed considerably since my day and with considerably more performance differed in behaviour near and on the ground, though easy enough in the air. As an engineer I complained that the *Osprey*, which was the naval version of the *Hawker Hart*, was in a poor state so far as its engine was concerned. I could hear the poor old thing knocking and felt certain it would not be long before it gave up the ghost. Anyway, my very kind and long suffering instructor let me put in time with the *Nimrod*, which was the naval version of the *Hawker Fury*—a single seat fighter. This was great fun and a lovely aircraft to fly, though at that time obsolete as an operational aircraft owing to lack of performance. It had a fixed, non-retractable, undercarriage and a biplane wing configuration. It was only in the last few days of the course that I got on to the *Miles Master*, a little aircraft arranged to give an idea to those flying it of the characteristics of the then modern fighter of the *Hurricane* or *Spitfire* type.

The *Master* had two seats so that an instructor could start one off and then get out when he thought you were competent to fly it alone. This was by no means difficult so long as you remembered the drill for operating undercarriage, variable pitch propeller, flaps, etc. Then finally came the real titbit, which was to fly the one and only *Gladiator* (see Fig. 6) on the station. It was only a single seat biplane but still just about operational, having a pretty good performance and in particular a very high degree of manoeuvrability which made it absolute bliss for aerobatics. For some years thereafter on various stations I managed to beg, borrow or steal one of these beautiful little aircraft, when feeling in the mood for some of the real pleasure which can be experienced in aerobatic flying.

Having finished the course at Eastleigh I was appointed to Arbroath Fleet Air Arm station—in other words HMS *Condor*. At that time the station was used for a variety of purposes, including observer training, and the services trials unit, No. 778 Squadron, was based at Arbroath. My appointment was as Chief Engineer of the station as a whole and also as technical officer and trials pilot to 778 Squadron. I was distinctly the dog's body so far as flying was concerned, as most of the pilots were very experienced and quite first class members of the Fleet Air Arm. We were commanded by James Tillard, who was unfortunately killed about two months after I left, along with two other most valuable officers, in a *Walrus* amphibian. The cause of this accident was never explained so far as I am aware, but it was a sad blow for the Fleet Air Arm.

They were all kind to me in the squadron and tolerant of my relative lack of experience as a pilot. I was allowed to fly all the aircraft available once I

Fig 37 Grumman *Martlet*

had flown the *Roc*, a naval fighter used at the beginning of the war up to about the time of the Norwegian campaign in the spring of 1940. This aircraft was rather underpowered for the purposes of a fighter and had the reputation of being a bit tricky near the ground. However, apart from the fact that it was obviously pretty heavy and had a considerable rate of sink when the engine was cut on landing, it did not seem too bad.

The one occasion when I might, perhaps, have been of some use to the squadron in my capacity as a pilot was when a squadron of Grumman *Martlets* were sent over from the U.S.A. on lease-lend and employed in a fighter squadron at Hatston in the Orkney's, their prime job being to protect the Fleet Base at Scapa Flow. In the hurry to get these aircraft operational they were sent direct to this squadron before evaluation by us in "778", which was our true function before aircraft were released to service squadrons. However, in view of the number of *Martlets* damaged due to swinging off the runway at Hatston, they were withdrawn from service temporarily, until our squadron at Arbroath had flown and reported on them. Accordingly all the pilots had a go on the aircraft, and so expert were they that no one even looked like swinging off the runway—until it came to my turn! Then slowly but surely as we came towards the end of our run I managed to let her go off to starboard into the mud alongside. Of course I should have given the engine a little burst to enable the rudder to become more effective, but instead I just sat there and looked at it! Anyway I suppose the squadron were able to report that there was some tendency to swing in these

Fig 38 The *Seafire*

aircraft and to give advice as to how to avoid the trouble. They certainly required a good deal of force applied by one leg to the rudder bar to take them off without swinging. I never saw the report, but it is perhaps enough to say these Grumman fighters were used extensively both ashore and afloat from carriers from this time onwards, until superseded by the *Hellcat* by the same firm almost eighteen months or two years later.

In the air they were delightful little aircraft to fly but it was a bore to have to wind up the undercarriage manually after take off. When in due course I came to fly the *Spitfire V* (called *Seafire*) I found the *Martlet's* controls were in comparison not very well matched as between ailerons and elevator (i.e. roll and pitch). In the *Seafire* the matching was beautiful. I feel certain this was largely due to the very good work done in the prototype trials stage by my very good friend Jeffrey Quill, who did the vast majority of the work of testing and developing the *Spitfire* for service use.

Probably the trouble with the *Martlet* on the ground was caused by a very narrow undercarriage and the fact that at the time we in Great Britain were more used to the type of brakes operated from the control column with the right hand. The U.S. type of brakes required the pedals on the rudder bar to be pushed forward by the pilot's feet. This, while quite satisfactory, took a lot of getting used to. When I actually came to fly the *Seafire*, after leaving Arbroath, I found it also a delight to fly in the air but a little formidable on the ground. She had a rather narrow undercarriage, and it was also difficult to see ahead while taking off or landing, the long engine cowling blinding the view for a little while on take off and approach. Deck landings were tricky, though I never had to do this myself. Of course the *Seafire* was a good deal faster than the *Martlet*.

The period spent at Arbroath passed all too quickly, it was such a wonderful short interlude from the shipyard, and was spent in delightful country, but I cannot pretend it was a particularly warlike interlude. The nearest I managed to getting any action with the enemy was due to the fact that for about my first two months up there the Battle of Britain was being fought, and though a good way north we had all the fighters belonging to the squadron armed, loaded, and ready for action in case an intruder appeared in the course of our flight. One particular day I well remember was when the padré came along in the afternoon and asked me to give him a flight in a *Fulmar*—the Fairey two seater fighter with the same Rolls Royce *Merlin* engine as was fitted to *Spitfire* and *Hurricane*. I asked permission to take him up from our C.O., James Tillard, who at the same time informed me there was a report from control at Petrevie of an enemy aircraft approaching from the sea in our direction. It was thought to be a *Junkers 88*. I was asked to go up in the *Fulmar* and try to engage it. In the general rush to get going I did not help matters by pressing the fire extinguisher button instead of the starter, which resulted in foam being squirted all over the engine, causing a delay of at least two minutes! We eventually got off the ground and climbed as quickly as we could, but saw nothing. Presumably the Junkers had passed over long ago, or had turned back after sighting the coast. Even had we seen

Fig 39 A *Fulmar*

it the Junkers was probably faster than us, so it was a fairly forlorn hope and we returned. When we landed it occurred to me I had never even told the padré we were up after an enemy aircraft; he was most surprised when I told him.

Other than a number of flights in the service aircraft of the 1940–1 period, my flying did not consist of anything sensational. One particular trial our squadron had to carry out early in my time at Arbroath was the flying of an *Albacore* TSR (Torpedo Spotter Reconnaissance) aircraft which had a Bristol *Taurus* sleeve valve radial engine that had been causing some alarm and despondency in operational service at the time. The trial was arranged so that we started with a new engine fresh down from the factory and flew it continuously in daylight hours until it actually failed. All pilots in 778 Squadron took turns of about three hours each and after close on a fortnight the poor thing became very ropey indeed. Each take off became something of an event with the engine vibrating alarmingly and throwing oil all over the windscreen, which made it difficult to see ahead. Eventually, of course, the engine gave up the ghost and a forced landing had to be made —luckily in the hands of a more skilled pilot than I. It was the impression at the time that the trials were a help to the engine designers—at any rate the *Albacores* continued to do much useful work at sea until aircraft with higher performance were introduced into the service at a later date.

I had a great friend in Harold Grenfell who was developing an airborne radar called ASV at that time (late 1940). As he was an observer he had to find a pilot, who would be one of our squadron (778), to fly him. We had many flights on Sunday mornings, partly perhaps to get out of church, but also because his development had the utmost priority. One particular Sunday there happened to be a thick bank of cloud at 2,000 feet and we went out to sea in about the position of the Bell Rock and after a short interval Harold told me they had picked up a convoy of ships running up the Eastern channel. I asked him what he wanted to do and we mutually agreed that we would attempt to surprise them by diving through the clouds about 4,000 feet thick. Much to my pleasure we came out nearly vertically over them, but my pleasure was somewhat marred by finding the convoy all training their guns on us with flags and Aldis lights or searchlights working at full chat. However, we had an Aldis light and soon made the necessary call sign so we did not actually get fired at. I have always been told this was the first time in the Royal Navy that an airborne attack had been carried out through thick cloud and come out spot on the target. Obviously this happened many times later in the war, but these were early days.

After about six months in the Fleet Air Arm the Vice Controller at the Admiralty and my chairman at Vosper, Admiral Sir George Chetwode, became insistent that I should return to the firm. The U.S. engines were arriving in the country and becoming available for installation in our MTB hulls in lieu of the Isotta marine engines. Accordingly the Fleet Air Arm were asked to release me, and I returned to Vosper at Portsmouth in about May 1941. I had seen for myself that things were getting into a bit of a pickle in the course of earlier visits in that year.

Fig 40 A *Walrus*

It was a blow leaving the Fleet Air Arm, which had been a wonderful change for me; though it would be absurb to say that I had done very much for the war effort in the short time I had been serving. However, by an arrangement with the aircraft ferrying squadron at Lee on Solent I was able to get in a bit of flying even after I had returned to Vosper, because there was a shortage of pilots available for ferrying service aircraft at that time. I used to specialise in delivering or testing *Spitfires* and *Walrus* aircraft and the occasional *Hurricane*. Strangely enough at that time there were few pilots about who had experience of the *Walrus*, the splendid Vickers amphibian. They were not difficult to fly but rather different from other aircraft, and quite fun to take off and land on the water. I got quite a lot of short hops in these Walrus amphibians mainly to Cowes in the Isle of Wight and to Halton near Aylesbury, where they specialised in their overhaul.

All this went on splendidly until a day came later in 1942 when I had to force land a *Hurricane* with a dead engine, having taken off from Yeovilton in Somerset for a test flight after overhaul. They had asked me to do the testing as quickly as possible because they were in a hurry to get her into service. I flew around for about 20 minutes then returned towards Yeovilton and was in fact making an approach to the field when the engine suddenly stopped. I was not within gliding distance of the airfield having only about 800 feet to spare. The first and most obvious thing to do was to change over to auxiliary fuel tank, but still there was no luck with the engine, so I could only try to make the best possible dead stick landing in the rather small fields surrounding Yeovilton. Thank goodness I managed to remember my instructions about keeping up flying speed in such circumstances and stuffed down the nose and levelled out when the level of the fields and their surrounding walls was reached. This was alright so far, but it was really impossible to get away with it completely in such small fields. I eventually had to stall her from about 8 feet into the best field I could see when the speed had dropped to near stalling point. Rather stupidly I had not fastened the additional harness to prevent my head going forward into the gun sight, otherwise I should probably have got away with it altogether. However, I was temporarily knocked out as the aircraft collapsed the undercarriage and went on her nose with the tail about 30° in the air. The attention I received at the sick bay, where they stitched up my forehead, was wonderful and the Surgeon Lieutenant who did the job told me I was lucky, because in private life he was a plstic surgeon. I must say I believe him, as after the scar disappeared very little marking was left showing on my eyebrow and forehead despite the stitches. I have often wondered where he is now—29 years later.

I was blamed, rather unfairly, because for a number of reasons which I will not go into here, I am sure the aircraft was quite empty of fuel in both tanks. However, in the hurry to take off the Flight Sergeant had assured me she was all ready and I signed the Form 700 which states that the pilot has personally checked everything including fuel. For this reason I had little to stand on, though I was very sore when they accused me of taking off on the auxiliary tank and not changing over to main tank.

Having flown so many types in 778 Squadron at Arbroath, this is a mistake I would never make, as in these circumstances one always suspects fuel first and changes over, or at least checks the position of cocks. However, that was that and as I was not really in the F.A.A. at the time I was stopped from flying—at any rate for the time being. It had been the best of fun most of the time and most interesting, even thrilling in a small way.

I was not, in fact, doing any harm and at times helped considerably; also there is no doubt in my mind that the claim by the station officers responsible was all rot. It was said by them that when they came to take off the filling caps from the main tanks they found them full. I asked the officer concerned and others whether any of this petrol flowed out as they took the cap off the tank, and unanimously they said "no". Well the aircraft was on its nose with the tail well up in the air at between 30° and 40°. I asked the makers whether this could be possible and immediately they told me that at least 200 gallons would flow out in these circumstances, since the filler orifices would allow a flow out before a level surface was reached. Besides which, one of the maintenance men involved took the trouble to come to the sick bay while I was there and tell me I was being taken for a ride. I could not use his name for obvious reasons.

All rather sad but they could always fall back on the fact I had signed the form 700, and I suppose they thought it better for me to take the rap as a semi-retired civilian than their service personnel. However, I could not really be surprised; it only showed the sort of thing which went on in wartime so far as maintenance was concerned. If only the next Captain, Conolly Abel Smith, had been in command I am certain he would have given me a square deal; and as a matter of fact he looked up the papers and told me it was all very fishy indeed. Conolly even told me I could have an occasional flight in his *Gladiator*, which was a real act of friendship. He afterwards became Vice Admiral Sir Conolly Abel Smith, KCVO having ended a great career as Admiral of the Royal Yacht *Britannia*.

Well, this is perhaps enough of my rather elementary but pleasurable air experiences at that time. There can be no doubt that I generated a completely new outlook on my job of designing MTBs, for I learned about fairly sophisticated instrumentation and other gadgets which was most useful.

8

Return to Vosper

We must now return to civilian life at Vosper. The firm had too much work to do and far too few highly qualified staff; in addition it had suffered badly from the bombing of Portsmouth, which continued for some time. Even if those in the target area were not frightened it was difficult to get any sleep. I did admire my secretary, Miss Carpenter, who refused point blank to leave her house right in the target area and was always controlled and efficient. I fear I worked her too hard, for my strong inclination was to go and deal with the hardware (boats), etc. for most of the morning and afternoon, then turn up about 6 o'clock to deal with my mail and dictate letters and reports. Very naughty and thoughtless of me I suppose but possibly understandable in the circumstances; we were so spread out with the offices in the country.

All managing directors have their best subjects and inclinations so tend to concentrate on them to the detriment of other duties, though we must hope not too much so. The answer, of course, is to get a damned good general manager working with his assisting managers of departments.

Our men were dedicated, but much of the work was spoilt by quarrelling, though they were individually as good as you could get for a relatively small firm. Some people could not see why I had to strengthen the staff from outside, and almost without exception the remainder closed the ranks against the unfortunate newcomer(s). This may have been reasonable in certain cases, but on the whole we got superior personnel when changes or additions were made. Things would have been far better if people had co-operated, but I cannot altogether blame them. The firm had grown from very little, and I should have taken more trouble to watch what was happening and have appointed a general manager under me earlier than I eventually did. Of course I could appreciate this, but having brought things up a long way since 1931 I did over much myself, walking round the shops and drawing offices and taking time in discussing the essential technical problems, which were many. One very great help to me was the return from the war in the Atlantic of Bobby Somerset, a great sailor, who had been with us before the war. Besides being a great personal friend he helped me restore order as Progress Manager, and was invaluable.

My mistake, I know now, was to keep the managers in their respective departments without sufficient overall coordination, which was probably all right when I was young and enthusiastic and possibly until the war; but when we got the mass of orders after the outbreak of war things became out of hand, and even foremen tended to go their own way disregarding the drawing offices or anyone else. Orders were placed from the foremen on their own initiative by telephone and seldom confirmed, so it can be imagined that the position became chaotic. The inventory of stores, spares, etc. soon represented an astronomical figure, totalling more than the rest of the cost of work in progress and completed. In addition we were very late on deliveries, which caused much trouble as little time was available for the quite inevitable teething problems which invariably arise.

One could say, of course, that I probably should not have gone into the Fleet Air Arm when I did in July 1940, but besides a lot of persuasion from friends in the Fleet Air Arm to the effect that I could be of great value to them at that time, being both an engineer and pilot, I found little to do at Vosper until the U.S. engines arrived. This was a direction in which the Fleet Air Arm was a bit weak at the time, having only recently taken over from the Royal Air Force all maintenance and engineering generally. In fact when I joined up there were still a number of R.A.F. Flight Sergeants serving in the R.A.F. uniform. The idea sounded interesting and attractive and I thought, in view of the failure in the supply of Isotta engines when Italy joined the war, that there would be little to do at Vosper for some time to come and until some alternative source of engines was found. Even then I thought the adaptation of the Isotta installation which had been worked out so carefully by Fred Street in conjunction with Lieut. Pratt, as he was at that time, that no problems would arise when we came to installing another type of engine. In this I was about 100 per cent wrong.

When the Isotta engine supply failed the Admiralty sent us some reduced power engines of U.S.A. origin and ordered us to instal them in some of the boat hulls which had been intended for the Isottas. This job, which, frankly, I thought would be quite easy, turned out to be a headache, as indeed did the installation of the Packard engines when they came along. It was unfortunate, though understandable, that Fred Street and Polly Pratt decided to instal these engines in the same manner, and using the same type of coolers, filters, salt water pumps, etc. that had been used when installing the Isotta engines. As there was a good deal of engine unreliability when these engines went into service in our boats, the Packard engineers, who had a free run of all the bases, the Admiralty and the building firms, took great pains to blame us for not following their detailed installation intructions. There was a good deal to be said on both sides. Actually the Packard engine, good as it eventually became, did want rather careful treatment at that time, much more so than the robust Isotta Fraschini. For instance, the oil flow through the engine had to be made greater than the piping and filters we had used with the Isottas were able to handle. It was not until I went to the U.S.A. to get the lease-lend MTB of Vosper type going that I realised there was a weakness in the casting housing the crankshaft bearings which allowed the crankshaft to whip and damage the bearings. This I found out by going over the repair shop at Packards in Detroit and finding that all engines in for repair had these bearings gone—even engines which had done only a small amount of running with the U.S. Navy who had, of course, installed them in accordance with Packard's requirements. This point alone was sufficient to account for much of the trouble we were experiencing with the operational boats.

This state of affairs lasted about a year while we got a lot of blame, which was most galling. It was, unfortunately, soon after the fall of France when the boats could have been very useful off the Low Countries, and many actions were being fought in these waters by our boats operating from the East Coast bases such as Yarmouth, Lowestoft, Felixstowe, Dover, and also from Newhaven, Portsmouth, Weymouth and Dartmouth. I had the mortification on one of my visits to be told by Peter Dickens at HMS *Beehive*, the base at Felixstowe Dock, that our flotilla of 40/41 class boats were known locally as Rotten Row, as they were all in a row made fast to the quay undergoing refit after breakdowns. This hurt me a lot as Peter Dickens was one of the really great MTB leaders at the time and had already made a great name for himself. I had first hand confirmation of it as I had only recently taken part in an operation with him in one of our boats off the Dutch Coast, when we tried to intercept a German convoy. As it happens we did not intercept it, being detected on our approach. We waited off the Hook of Holland till daytime, then returned.

This spurred us on to greater efforts to make all the changes necessary by sending a lot of our men to Felixstowe, and gradually all became well again. It has to be said in fairness to ourselves that subsequently the Admiralty found it necessary to adopt our salt water pump in place of the original scooping arrangement; but the Packard eventually turned out to be a thoroughly good and reliable engine. This applied also to their 1,500 BHP engine. We heard from the Swedes they had considerable teething troubles with their 1,500 HP version of the Isotta which they bought during the war for their MTB flotillas.

The production organisation of these little fast boats belonged to the Admiralty in the final resort. It became very difficult and to some extent unpleasant, for we had the worst of both worlds in being responsible for the design, which I am sure some elements of the Admiralty did not like at all, the inference being that they had not sufficient experience for the job. It all left a lot on our plate and we had to get on without delay, ordering where we could to get the boats out in time. Probably we were frustrated and argumentative, but we genuinely thought we were most likely to be right. Of course, as the war proceeded things became much better, but one had pretty frequently to go up to the Admiralty to see the Vice Controller,

Admiral "Tommy" Tower, who was in charge of small craft as a whole. Many times I found that the naval construction (D.N.C.) section of the Admiralty would give way to the "Staff Requirement" as formulated by the operational officers. They, for very good reasons, would want something near to twice the armament and other essentially warlike equipment that could be carried by the boat concerned if she were to retain a worthwhile performance, or any performance at all in view of the difficulty of even planing when overloaded. On numerous occasions I had to do what I could —as tactfully as I could I had to refuse to fit certain equipment so as to avoid the inevitable lack of performance when the boats became operational. Incidentally, almost invariably we "took the can back" if any shortcomings were found by the operational officers. One understood the thinking of the young officers, but it was grossly unfair on us in most cases.

One of the worst examples occurred when a shipyard in the Isle of Wight found some fairly high powered engines in the U.S.A. They took a licence to handle the agency in England, then went up to the Admiralty and suggested that the engines should be installed in some of our Vosper 1940–1 class boats. We were not consulted at any stage. The weights had been very carefully worked out by us for the Packards or Isottas and there was little enough margin. Any gross departure would result in the boat being hopelessly out of trim, and hence painfully slow and useless as a fighting ship. So far as I recollect the difference in weight aft was about 4 tons. It is almost impossible to believe that I had never even heard about the change of engine; which, however did not inhibit the various officers concerned from heaping abuse on me, personally. I even became unpopular among some of my own friends! Obviously in order to please some rather insistent shipyard representative, this silly piece of work was put into effect by the Admiralty without consulting us. It was extremely unsatisfactory, but having got along fairly well on our own up till the war and some time after, and having produced what was admittedly a good boat with the Isotta engines, we found ourselves in the unenviable position of carrying the responsibility and blame, though so much had been taken out of our hands. The really galling part was that we had paid £15,000 so that this good engine could be built in England, and this was quite wasted. A severe gap in the production of satisfactory boats happened at the worst possible time.

If ever there is another war calling for cooperation between rival firms some way must be found to eliminate continual jockeying for position amongst them. The reader will be inclined to ask why the Admiralty or equivalent department should not make the design, which would be sent out to the various firms to be built to specification. It sounds fine, but in fact this type of high performance small craft lends itself much better to the type of production used by the aircraft industry, or aero space as it is called nowadays. Both the *Spitfire* and *Hurricane* were designed by private firms in the first instance, as were all operational aircraft then and still today, so far as I know.

The design staff of a good firm is an almost irreplaceable asset, taking the best part of a generation to build up. Design is also a highly individualistic process in which one man must be clearly in charge of every stage, and on to whom the responsibility falls in the event of unsatisfactory performance. Clearly, there must be cases where such a chief designer cannot, in fact, be responsible for the detail design of all units, but it is very much up to him to make certain that the components incorporated in the aircraft or fast boat are capable of giving their performance reliably.

No one expects the aircraft designer, for instance, to design the details of the engine units, but he has to see they are capable of supplying the necessary thrust to enable the aircraft to give the required performance in service; also he has to ensure that the engine unit is given a chance to perform satisfactorily in the aircraft by making certain the installation and supply of fuel, oil and control services have been carried out to his satisfaction. This particular task is not straightforward and, combined with the starting sequences, involves an engineering study of extreme complexity making use of many electronic techniques; this is especially the case with gas turbines, where the sequential starting functions commence by turning over the compressor and end by supplying and flashing up the fuel supplied so as to produce the necessary idling performance, whether it be in the form of a jet efflux or a revolving free power turbine.

In 1943 a committee was set up by the Admiralty with the object of developing an engine suitable for MTBs or Fast Patrol Boats of the future.

One of the more obvious lessons gained the hard way during the war was the absolute necessity to develop suitable machinery for these craft some years before any warfare is likely to break out. It was obvious even at this stage that the boats were likely to become bigger as time passed. There are a number of reasons, the more obvious being superior seakeeping, combined with the ability to carry a sufficiently powerful weapons system to render an effective attack possible against larger vessels; and also defence against any fast patrol boat types likely to attack in defence of the larger ship.

The commission was placed under the chairmanship of Sir Roy Fedden, a well known engine designer who for many years had led the design team at the Bristol Aircraft engine company.

Most of the firms engaged in designing and building fast craft were asked to appear before the committee and to offer their opinions or advice. It was obvious that the bias was heavily loaded in favour of a diesel engine, though personally I was in favour of developing the gas turbine which had recently proved successful in aircraft.

I discussed with Roy Fedden and his committee the possible use of gas turbines for MTBs and similar craft. Fedden played down the possibility at that time (1943). He was very much in favour of high speed diesels offering about 3,000 BHP per unit. This afterwards became the Deltic engine, which has done well on the whole but was mainly suited to the slower round form craft; though it was fitted in the *Dark* and *Nasty* class of planing boats.

9

Wartime Visits to the U.S.A.

In July 1941, after the U.S.A. had embarked on a programme of lease-lend, I was sent over to help get the programme started by the Annapolis Boat Co., who had negotiated a licence with Vosper.

Although the U.S. Government would not allow royalty payments to be made to any person or firm out of the funds provided for their armaments and supplies made under lease-lend, it expressed itself as being ready and willing to allocate the production of suitable firms, with the object of producing the Vosper type MTB in the U.S.A. for the use of the British Navy.

In view of his enterprise in negotiating a licence agreement with Vosper Mr. Chris Nelson, the managing director of the Annapolis shipyard, was detailed by the U.S. Bureau of Ships to act as parent firm to the remaining boat builders selected by me in the course of visits to likely yards that seemed capable of undertaking such relatively sophisticated work. In fact, the U.S. Bureau of Ships supplied me with a pass to travel on any airline and asked me to visit a number of shipyards recommended by it. It also gave me permission to put forward the names of any yard I should come across that seemed likely to be able to do the job satisfactorily.

The only shipyard I found not on my list was the Harbour Boat Co. of Los Angeles, run by two very capable Italian immigrants who were mostly occupied at that time in the building of tuna fishing boats. The yard proved most efficient and to the best of my recollection produced MTBs of Vosper design to as high standards as any yard in the U.S.A. or for that matter in Great Britain.

It must, of course, be borne in mind that many of the more complicated and sophisticated installed items, such as main and auxiliary machinery, shafting, vee drives and propellers, were ordered from specialist firms to meet the same specification as a similar item installed in Great Britain. The whole electrical installation was fabricated in U.S.A. to British specifications and distributed to the various yards for installation. The only items actually supplied from Great Britain to the best of my recollection were the armament and ammunition besides possibly—though I am not certain—the torpedoes, tubes, and firing gear.

The journey out to U.S.A. at that time was in itself a minor adventure and extremely interesting. For security purposes when dealing with the U.S. Bureau of Ships, which would normally look askance at a civilian, I was sent out as a Naval Officer having just left the Royal Navy (Fleet Air Arm). However, at that time it was difficult to get across the Atlantic by air from east to west, without landing for refuelling at Lisbon, the Azores and Bermuda, whence one could fly in one hop to New York. In order to reach Lisbon a civilian passport was provided and all officers were instructed to travel in civilian clothes and without rank.

I was instructed to report at an airfield in Bristol early in the morning, which necessitated travelling to Bristol the evening before and staying in an hotel. The airfield in Bristol was small; the aircraft a DC3 twin engine machine belonging to the Dutch airline KLM. The aircraft and their crews were operating from English bases, having managed to escape from the Germans, who by this time had overrun their country.

The windows in the passenger compartments were covered with shutters so that we could see nothing outside. However, after about 20 minutes flight, in what was obviously a westerly direction we landed at an airfield which was not named to us but which I recognised from my Fleet Air Arm days as Chivenor in North Devon. Here a full load of fuel was taken aboard, and we took off in what was fairly easy to see was a northerly direction as we were shortly flying over the sea.

The exact route followed to arrive at Lisbon 7 hours later was not disclosed to us but it must have taken us well out into the Atlantic so as to avoid, so far as possible, the attentions of long range German fighters; only the week before they had attacked and shot down the aircraft on this flight, when that superb actor Leslie Howard lost his life. It was difficult not to be anxious, but compared to the risks taken by an operational bomber's crew over enemy territory one felt stupid for giving the possibilities even a moment's thought. On arrival at the Lisbon airfield after an eventless flight we found, to our surprise, a *Junkers* passenger aircraft landing from German territory and taxi-ing up to the same airport buildings, where we were all mixed up together—friend and foe—claiming our baggage and producing documents for Portugal's neutral officialdom.

We discovered there was a wait of 2 days before the Pan American Clipper flying boat would proceed across the Atlantic. My hotel was allocated by officials operating from the British Embassy and was the Grand Hotel at Estoril, a fashionable watering place some 15 miles out of Lisbon on the coast. It so happened that I knew a fair number of the Portuguese residents, as we had stayed there several times before the war. As always these friends were hospitality itself, and it is not difficult to understand the effect on one used to the blackout and food shortages in England.

All of us passengers making for U.S.A.—there were not many—were warned that Lisbon and environs were a hotbed of espionage and to be very careful of documents or giving information, if one knew any, which could be of use to the enemy.

My only contact of this sort of thing took place when a woman who looked like a school mistress waved at my taxi driver on the road from Lisbon to Estoril. On stopping and, quite without reference to me, the lady jumped into the back of the car with me and within two minutes had stripped herself to the waist leaving nothing to the imagination as to what was expected of me. My embarrassment can be imagined when we came to a railway level crossing whose gates were shut and we found ourselves alongside an important lady of the English Colony whose face expressed a mixture of disapproval and disgust. It was embarrassing as I knew her fairly well from pre-war days!

As we got near to my hotel I told the driver to stop and ask the lady to step out. She did so, somewhat reluctantly, and I asked the driver what it was all about. He said she was well known for this type of behaviour and came from a local nut house or the equivalent, only being allowed out in the day time. I suspect this was scarcely the truth and that if my reactions had been different the driver would have shared the spoils with the lady concerned.

Our voyage on from Lisbon to the Azores and Bermuda was without much incident and as the U.S.A. was not then at war, without much danger. There was a row about which I took an innocent and watching part when a man who turned out to be the German Consul designate for one of the South American countries demanded from the captain, who was, ship like, on a deck above the passengers, that I should be removed from the cabin where his wife and two daughters were, as at night it was really only curtained off from the rest. I forget the outcome; so far as I recall I gave up my bunk to the gentleman concerned. In any case the *fraus* lacked attraction so the surrender was no loss.

It was, to say the least, peculiar to find a number of passengers joining at Bermuda who were obviously dirty old men—and some relatively young—accompanied by what are usually referred to in the U.S. as "high steppers". Their gaiety and flamboyant clothes were such a contrast with anything we had seen in England during the last few years; so too was the easy and pleasurable way of passing the time, which to us was always pressing in those bad days.

We passed through some rather nasty turbulance caused by an inter tropical "front" and besides squeaks and clutchings from the nearest man, which happened to be me, two of the very smartly dressed little girls passed clean out until about an hour after the storm was over. They were an extraordinary sight. I had never before seen eyes showing only the whites—no pupil was visible. The captain, who had left the controls after passing through the storm, asked me if they had been hit by luggage or heavy objects falling from overhead. I told him I thought they were in a shocked state, as for the first 5 minutes they had been handling their beads, and appeared from what they were saying, to be confessing their sins; which were plenty for girls barely out of their teens.

On arrival in the New York flying boat base I was met by Chris Nelson, who looked after me so well that I had to retire early and sleep off the effects of the trans-Atlantic flight and his hospitality. I slept about 15 hours, which is a record for me.

On arrival at Annapolis I was introduced to the leading personalities and we got down to a long discussion right away. There did not seem to be much difficulty about the hull at this stage, but such things as vee drives, castings for stern tubes, propellers, etc. were an obvious cause for some concern until we had found the right sub-contractors. The U.S. Bureau of Ships was of infinite help here because it had a very close idea of what firms might be able to undertake the job required. There was a good deal of discussion about the electrical installation, mainly because of the very tight specification required by our Admiralty in England.

The first visit to lay the foundations of the programme and to satisfy myself that the Annapolis shipyard had enough to get well away lasted about three months. They passed quickly; but the meetings in Washington were pretty frequent and involved much discomfort and anxiety, as all accommodation had to be allocated officially by the British Purchasing Commission, who seemed to be able to get nothing better than small rooms in the roof without air conditioning. Even then we would not know to what hotel we had been allocated until about 6 o'clock, when the room clerk, who I imagine made little profit out of our small dollar allowances, would announce with some emphasis that the booking lasted for one night only. So the dreary business would start again next day, while we attended endless conferences and another unwilling hotel was allocated to us if we were lucky. If we were out of luck I would usually drive down to Annapolis and come up in the morning. The months of June, July, August and September are particularly trying in Washington, as the temperature is nearly always in the mid nineties with humidity never far from 100 per cent. One of our party, succumbed to the heat and general discomfort. Poor chap he never slept a wink and became more and more depressed, to an extent that I had to ask the British Naval Attaché to do something for him. He was taken into Bethesda, the big naval hospital outside Washington, where I don't doubt they applied shock treatment and other forms of therapy. The poor fellow was shipped home and the last I heard of him was that he had escaped from a psychiatric hospital in England and had thrown himself under a train.

When I had finished as much as I thought I could do in U.S.A. I volunteered to fly home, a bomber placed at our disposal under lease-lend. On reporting at Dorval in Montreal I found that I should have to do at least two weeks conversion course on to twin engined aircraft. The Fleet Air Arm, in which I had qualified, were at that time all single engined aircraft and the *Hudson* which we were asked to fly was not of the easiest aircraft to handle. So they allocated me to an aircraft having a very experienced U.S. civilian pilot. I don't doubt I could have got the hang of it in a week, but it seemed a waste of time, as I had already been away for 3 months.

We flew without incident to Gander in Newfoundland and then slept in tents for a few nights before the weather in the Atlantic was considered suitable for 25 of us to take off from the very long runway, well lit, on which we were told to keep the aircraft down until 120 knots was reached, when we could "rotate", i.e. lift off. All went well, but the little Hudson was full of petrol in wings, fuselage and bomb bay in order to reach Prestwick in the morning. Our crew consisted of the U.S. pilot, myself as second pilot, an R.A.F. sergeant just trained in Canada as navigator and a Canadian civilian as radio operator. Due to lack of radio communication, which was the result of exceptionally strong Northern Lights (Aurora Borealis) we found ourselves flying almost permanently in cloud at 7,000 feet. If we had been able to communicate we should have received the radio order sent out from Gander telling us to climb to 10,000 feet. As it was the navigator had no chance to get a star sight and so check our position; it was 5 hours before this became possible, when we decided to take a chance on it and climb out of the cloud. Our man, just trained and on his first operational flight, took his star sights and then spent a whole hour working them out while we were saying to one another—"can he or can't he?" However he did eventually give us a reasonable position on the basis of three stars and we altered slightly to the north. One does not want to do too much altering to the north in a case like this, as it is all too easy to miss Scotland and end up short of fuel somewhere between Wick and Bergen.

However, after about 9 hours we sighted Ireland just about at Galway Bay and skirted the coast towards the good airfield at Londonderry. I was all for coming down there, refuelling and getting a weather report before continuing to Prestwick, but the U.S. pilot thought he could make Prestwick all right and set a course which took us into some nasty wet, dark, woolly clouds. We climbed above them and looked around for a hole through which to spiral down to get near the surface and under the cloud.

We succeeded in getting under the clouds, but by this time we were so near the surface of the water that we found ourselves below the mast tops of some ships and the Isle of Arran. Also, worse still, below a number of balloons which indicated the region of Greenock; in other words we were too far to the north. The U.S. pilot who was at the controls said he could make little sense of this dense mist and rain and asked me to do what I could to get a radio bearing from Prestwick, which was exactly what about 20 other Hudson aircraft were doing. However, to do the thing properly I called the distress signal Mayday and asked for a QDM (bearing). This brought a response and shortly afterwards we sighted the airfield.

Of course after we had landed we were told there was a perfectly good homing beacon at work at Prestwick but no one told us this at Gander and no one on board knew of it. This would have saved us some anxious moments, but I still think it would have been more sensible to come down in Northern Ireland where the weather was clear and good.

For those with an interest in technicalities, I should mention we had some anxiety while flying in cloud about 3 hours out from Gander. It is, perhaps, of special interest to those who still fly piston engined aircraft. We noticed that the boost gauge on one engine was reading substantially less than on the other despite the throttle opening being the same. We tried more carburettor intake heat, thinking that we were suffering from carburettor icing, which would not have been at all impossible in October in those northern latitudes. However, this made no difference; so rather worried we carried on and made up our minds that we would get out of these clouds as soon as possible. Then some bright spark noticed that the two speed supercharger controls were in different positions. On returning the lever of the one which read low on the gauge to high speed, the boost levelled off and all was well in the cloud for some more hours. During this time the radio operator reported that one of our number had got into a spin and lost 5,000 feet in height.

Most of these aircraft—unlike us—had only one pilot and although the auto-pilot was used for most of the time in this particular type the fore and aft trim had to be controlled manually. What had occurred was that the one and only pilot had left his seat to attend to the wants of nature and had inadvertently caused the nose to rise to the extent that the aircraft stalled and spun before he could get back to the controls. Luckily he was able to get her out of the spin and turned up bright and gay the next morning; this, amongst others, is a very good reason for having two pilots in each aircraft on a long flight such as this. Also it allows the pilot not at the controls to carry out a lot of monitoring, watching the navigation, keeping in touch with the radio operator and maintaining a look out for sight of the surface or runway lights while the pilot, at the controls, has all his concentration on the artificial horizon and other instruments. There is quite a lot to be said for some division of responsibility of this nature in the modern airliner when making an instrument approach to an airport in low visibility. I am not sufficiently up to date to know how this responsibility is divided in the jet age, though in most cases where conditions are marginal it must be assumed the captain will carry out the final approach and touch down on the runway.

Of the 25 Hudsons which started at Gander the previous night only 23 were seen again. The truth of the matter was that the fuel carried, even with the extra tanks fitted, left too small a margin for navigational error or unexpected head winds. Shortly after our trip the direct route was changed, with landings at Goose Bay and Iceland on the way across; though longer, this route was safer.

My second visit to the U.S.A. was carried out in 1942 at the request of the British Purchasing Commission, when the first Annapolis boat was reported as almost ready to run trials. This time I was sent over in a convoy of fairly fast passenger ships from the Clyde to Halifax; this took about seven days and was quite pleasant.

I well remember the day we took the first boat out in a thunder-storm with no decking over the machinery space. So far as one could tell the machinery was operating satisfactorily and I left them to get the boat ready for running official trials.

To return to England I had the good fortune this time to meet up with Lord Louis Mountbatten who was over in Washington mainly, I gathered, to sow the seeds for the eventual re-entry into France. One of the most important of his functions as newly appointed Chief of Combined Operations was to order the necessary craft for landing on the enemy beaches. Lord Louis, or Dickie as his close friends called him, agreed to give me a passage in the Liberator aircraft in which he was going direct to Prestwick from Montreal. His only stipulation was that I should act as his A.D.C. from Washington through New York and at Montreal, where he particularly wanted to see only the Chiefs of Staff and not to be bothered by the numerous other persons, of perhaps minor importance, at this particular juncture.

All went well in the train which was to take us to a small station just outside Montreal until, quite unexpected by me, I found the train stopping while I was still in bed and hardly awake. In fact I had to gather my clothes together, throw them out of the window and follow in my pyjamas to prevent being taken on in the train. I shall always remember Lord Louis' face of disgust combined with just a little amusement when he saw his "aide" in pyjamas of somewhat exotic colouring holding up his trousers with one hand and gathering his clothes with the other.

He went off to his meeting while the rest of us drove to the airport, where we were told that the Atlantic weather was quite unsuitable for any crossing that night. This, however, was reckoning without Air Commodore Powell, Chief of this section of Transport Command, who said he was quite happy to fly through any number of storms at about 8,000 feet provided His Lordship did not mind being bucketed about a bit. We were told, though I don't know how much truth there was in it, that the Air Commodore would fly through anything that came his way in the matter of storms but would not go high, although we were equipped with oxygen, because he suffered from high blood pressure. We got there after 23 hours in the air in special warm suits and with our oxygen unused. There were some nasty patches of icing and storm areas to traverse during the night, which gave the pilots plenty of work on the controls and doubtless some anxious moments. I watched them struggling to maintain a reasonably even keel—to use a nautical expression—and I noticed both pilots working hard at the controls in the worst patches of weather, but they would not trust me to give them a spell of relief while they were on manual control, though while on the auto-pilot in smoother conditions they were happy enough to let me fly her from the right hand seat.

Towards the end of November 1942 I was asked to go over and witness the official trials of the prototype Annapolis MTB. This time I was transported in the *Queen Elizabeth*, which was being used as a troop carrier and was more or less immune to submarine attack by virtue of her very high speed. We ended up at Halifax, and as senior of the naval officers on board

I was put in charge of the naval contingent taking the train to New York. It was no sinecure as every time we stopped at a station half the naval ratings would make a dash off the platform to the nearest public bar. Luckily for us it was a Sunday, when no liquor was served in this part of Canada.

The stations, unlike those in England, had no ticket barrier so it was all too easy. In order to get the ratings back to the train the engine driver would first of all blow his whistle several times and then start to puff out of the station. This technique worked until our arrival at Montreal, where an hour's stop was scheduled. Here the situation got quite out of hand and about 25 men never returned. On arrival in New York I thought there would be much trouble, but the officials there were used to this sort of thing and I heard no more about it—probably a number became deserters who were never seen or heard of again.

On arrival at Annapolis I found the shipyard a hive of activity with a charming young man called Dick Miller representing the U.S. Bureau of Ships. There was also a number of British naval officers who were standing by to operate and deliver the MTBs by sea to New York. One of them I remember well, Hilary Gamble, who had been awared the Distinguished Service Cross for his attack in the rather under powered MTBs stationed at Dover when the *Scharnhorst* came through the Channel in her extraordinary dash from Brest to a German port at Wilhelmshaven.

To refer for a moment to this extraordinary feat on the part of the Germans and equally extraordinary muddle on the part of the British forces, it is not generally known that a flotilla of the best Vosper MTBs, which with their Isotta engines could achieve more than 40 knots in operational trim, were sent down from their base at Felixstowe to Dover two days before the escape of the German ships, and every one of them was piled up on one of the many sandbanks in the Thames Estuary. They were therefore out of the fight, which otherwise might possibly have resulted differently; for they all had the speed necessary to get into position to fire their torpedoes without interception by the slower "E" boats, which in those days were incapable of much more than 30 knots. However, even with their superior speed they would not have found an attack all that easy owing to the exceptionally heavy defensive armament concentrated by the enemy in the Dover Straits area, and indeed throughout their passage to Germany. But skilfully handled as they would have been—all these boats had claimed a number of successes against enemy vessels—it is just possible they might have scored some hits with their torpedo armament, and at the same time drawn the defence from the unfortunate *Swordfish* attack by the gallant Esmonde and his squadron, and the equally gallant MTB attack by Pumphrey in command of the slower MTBs.

To return to the U.S. production: I found the first boat pretty nearly ready to run, but we were bedevilled with ice formation, which was a frequent hindrance to running trials in those upper reaches of Chesapeake Bay. When the day came for the official trials Richard Moss, representing the British Director of Naval Construction, Dick Millar and others were present, and to our surprise a speed two knots in excess of what we had obtained in the U.K. was achieved at the same load and with the same power. This was most interesting and satisfactory to the U.S. element including, of course, Chris Nelson and his boys at the Annapolis shipyard. For me it posed a problem, as there was no obvious reason for the improved performance.

After some weeks of close investigation and measurement of the hull I eventually traced the mystery to the fact that the draft markings both forward and aft were secured in the wrong positions on the hull, giving a false reading for the displacement, which further investigation revealed to be some tons lighter than the British boats. Hence the increased speed. I do not for a moment think the mistake was made on purpose. In the best shipyards the world over this source of error is commoner than most of us realise.

When delving closely into such matters it is surprising how often it can be proved that a hull has been built differently in quite important respects from the designed form. When we tried to use a standard electrical system making use of the pre-fabricated Plessey system, we found to our discomfiture that in more than one case a bulkhead would turn out to be located one inch out of place this way or that, which would have a disastrous effect on braided cable cut to a standard length.

The results achieved by the U.S. built Vosper type boat were quite good enough for practical purposes and many good little fighting ships were turned out from the yard at Annapolis and its sub-contractors, which included Harbor Boat Building, Jacobs and Herreshoff.

On the fourth occasion of visiting the U.S.A. during the war I found the Annapolis yard in full swing with its production and my presence not of much help. As the war was obviously approaching an end I took the opportunity to look around the industrial scenes in the U.S.A. with the idea of finding some product, technique or diversification which would help us maintain our production when the war orders were terminated.

The two best bets, so far as I could see in the few days at my disposal, were first of all the development and manufacture of deep freeze technique, which was catching on very quickly in the U.S.A. but was almost unheard of at that time in Great Britain. (This does not apply today when many quite modest establishments have their deep freeze apparatus.) Here we probably missed getting in on the ground floor, but it is possible that our resources were hardly up to the enormous development, which was even then being studied by such mammoth organisations as the Leverhulme interests and I believe also by Lyons and possibly ICI or one of their offshoots. Second of the bets was an endeavour to obtain the agency for Beech Aircraft, which, to my first hand knowledge were an excellent organisation for the production of what is still today one of the best of the smaller privately owned aircraft.

Possibly, in time this would have eventually proved a winner as it cer-

Fig 41 *Gipsy Moth*

Fig 42 A Beechcraft *Traveller*

tainly is today, 20 or more years later. We made use of privately owned aircraft at Vosper for a few years after the war, and very useful they were on occasions and until the airlines, both internal and intercontinental started to operate regular services. The drawback I found in running a private aircraft in those late forties and early fifties, was the difficulty in finding a really reliable radio for such small aircraft and the attendant difficulty in obtaining reliable and reasonably priced maintenance facilities. However, we owned first a *Gipsy Moth* (Fig. 41) then a *Proctor*, which did us very well and I took it as far afield as Rome; then we purchased a *Beech Traveller* from the Royal Navy, who had been using them for communications purposes between their various airfields. The latter was an excellent aircraft with a Pratt & Whitney Wasp junior, one of the best engines of its size available at that time. The Beech was fast and easy to fly in the air but not quite so easy to land on a runway with a side wind, the undercarriage being spongy and narrow although retractable in the air. This was a fine aircraft, but too complicated in its electrics for the maintenance available in the Portsmouth and Southampton area.

If we could have obtained the Beech licence and held out for a few years I believe it would have become a profitable concern, although we, as light fast boat builders, had but little idea of the special techniques required for aircraft building or repairs, let alone maintenance. Neither of these ideas were put into operation by Vosper in England, which may have been a pity.

10

Some Technical Lessons

After the war we, at Vosper, were pretty hard up for work excepting repairs and refits, so I thought it necessary to look round the world for work. One of my first voyages consisted in going out to the Argentine where the Isotta people told me the Navy wanted us to set up a joint factory to build our MTBs with Isotta engines in them. This sounded good and the whole idea might have been fine except that on arrival in Buenos Aires I was informed that Signor Caproni, representing Isotta, had been detained in his country. Where was I without the engine builder; but our agents were live wires and put me on to the Navy who said they wanted a number of patrol boats. On hearing this I cabled to Vosper in England and bought 12 "B" type Fairmile hulls from the disposals board, as well as 50 GM6–71 diesel engines, which we had to convert to marine engines by fitting gear boxes with reverse and clutch. With two of these engines installed giving about 160 HP each these craft made good 12 to 15 knot patrol boats and our agents assured us we had got the job.

Meanwhile I had enjoyed polo in the Argentine, which is like tennis anywhere else. I had played a good deal in Malta with the Navy and also at Aldershot when I was in HMS *Malaya* refitting at Portsmouth. It is a marvellous game when you can get someone to organise it for you, but having come out to sell boats I had to borrow boots, breeches and polo sticks. Count Romano, whom I had met in England, was the generous provider of polo ponies and good ones at that.

Travelling via Chile, Peru, Jamaica and Bermuda, where I stayed the inside of a week, I returned to England, full of the joys of spring with the good order.

In order to get to Chile it had been necessary to cross the Andes by air, and as it was 1946 we could only fly in the Panagra DC4, which is not pressurized. The first time we tried to get over we flew into a thunder-storm and after bumping about for about an hour returned to Buenos Aires. The next morning we had another shot and got over in good weather. I was immediately taken in hand on arrival at Santiago by Augustine Edwards, who handled our agency in Chile; he had a charming household with part Spanish and part English descendants. Chile is an extraordinarily attractive place with a good climate at Santiago and Valparaiso near the coast, while within one hour in a car one can be on some of the best skiing slopes in the world.

I had agreed somewhat reluctantly to be flown by Augustine Edwards in his Beech aircraft to see his skiing lodge, which you can imagine was well up in the Andes. To be honest I had had enough of flying after two attempts to cross the Andes at 25,000 feet and sucking oxygen in the cabin instead of dry martinis, which were more in my line. It was none the less a pleasant flight and the owner a good enough pilot, though I did not really appreciate his dive bombing methods of showing me his skiing lodge.

On return to Santiago we found a game of cricket being played on the airfield!

The rest of my happy journey to England up the west coast of South America was flown in a South American Airways *Lancastrian*, which was almost literally a *Lancaster* bomber with a few wicker work chairs and a very pretty hostess. The four Merlin engines made an awful noise but the plane was empty except for myself and a Dane who had been 11 years in Peru and was returning to Denmark, having made his fortune. He was celebrating the whole way home and got tight at every stopping place. The final effort occurred at Kingston Jamaica where rum punch was served, on the house, to all passengers arriving. This was too much for our friend. He was, in a short while, insensible—quite unconscious—so that the crew of the *Lancastrian* and I had to carry him to his room in the local hotel.

On arrival back in England, having had a thoroughly good time in Bermuda, I found the Argentine Navy had placed their order elsewhere so there was I stuck with 12 hulls and 50 engines. One of these hulls I con-

Fig 43 Sea Victory at Du-Clair on the River Seine

verted to a yacht with two of the GM6–71 engines and the *Sea Victory*, as she came to be known, gave myself and family nearly 20 years of happiness. She was a real home from home. I justified her existence by carrying out all the early development of the Vosper roll damping in her (see Appendix iii). Another became Loel Guinness' *Sea Huntress* which I write about below.

It was obvious we should have to go through a most difficult time at the conclusion of hostilities, but we had by this time amassed a vast amount of know-how concerning the design, running and building of the high speed craft of the MGB, MTB or Air/Sea Rescue types. I felt strongly that if we could hold out for say four or five years by undertaking repairs to wartime craft within our capacity, and not of necessity MTBs, we should be in an excellent position to take on further design and construction work, using the interim period to study, in depth, the lessons which had come to us, most forcibly in some cases, from the operational flotillas engaged in close combat with the enemy.

While the Admiralty received reports and suggestions, and quite a lot of criticism of the design and functioning of those little fighting vehicles, we at Vosper made a point of keeping in contact with the various flotillas based on the east and south coasts of England, where the most intense fighting took place from 1940 onwards, until the Normandy landings in 1944.

I have also described how, when occasion offered, I accompanied operations, usually aboard the senior officer's boat, which certainly gave a very worthwhile idea as to the fighting qualities of the craft at first hand. It also made clearer than at second, third, or fourth hand the limitations complained of. After the first few boats fitted with Isotta engines, which had on the whole acquitted themselves very satisfactorily in every way, we were much handicapped, as recounted above, by the necessity, at first, of using American engines of much reduced power. This greatly reduced the fighting potential of the boats we had to deliver to the Royal Navy. It was also found that an otherwise good hydraulic steering gear proved to be excessively vulnerable to enemy fire, so much alarm and despondency was sometimes caused by the loss of steering control in the middle of an action.

A mass of information had been gleaned during these war years, and soon after hostilities I asked the authorities to be allowed to design one of the 1944 class of boats to incorporate all the lessons learned, including the design of improved hull form and construction. The boat was approved and called *MGB538*, later changed to *1601*. The modified hull form was used until quite recently, over a space of 20 years, which included the *Brave* class FPB, the 68 foot Air/Sea Rescue craft and the second *Swordfish*.

There were a number of other modifications and novel methods incorporated, some of which were a success and others not so. The construction, while still all wood, made use of plywood techniques and the joints were glued in almost all cases, thereby eliminating the stress raising effect of the usual type of fastening, whether it be nail or bolt.

One experiment which, if successful, could have been useful consisted of combining in one generator a 24v. D.C. armature with an alternator on the same shaft, whose output matched the many electronic, radio and radar sets which in boats built up to date had invariably involved a number of converters, usually of the motor generator type, and thereby had caused additional weight, complication and expense.

While MTB1601 was being designed and built the Superintendent at the Haslar Tank, Dr. Gawn, who had done most of the propeller designing for our boats, obtained authority to have a number of different propellers tested on our boat MTB524 of the 1944 class. These trials, which were carried out by our own crews, included propellers submitted by Mr. George Selman, who had designed propellers for The British Power Boat Co., our closest rivals.

The latest boats produced by Vosper had two wing engines driving propellers at shaft speed and one centre engine with a 1.5 to 1 reduction gear effected in the vee drive. This propeller was two bladed instead of having the usual three blades, which are easier to balance dynamically. However, the optimum configuration eventually turned out to be a three shaft arrangement with 1.5 to 1 reduction gear on each engine, the two wing gear boxes being of the straight through types. George Selman had produced propellers running at engine speed with cambered face and back. This configuration was very successful but did not quite reach the Haslar two bladed reduction gear arrangement.

There is little straightforward in the design and construction of these little fast craft, and even after so much wartime experience unexpected difficulties would appear, often due to the increased weapon load the boats were being asked to carry. As with aircraft, in particular, and even in motor vehicles, tanks, and in the present day hovercraft, the user will nearly always ask the designer to meet requirements which are beyond the capacity of the available machinery, materials, or the existing state of the art.

I was often concerned when the Vosper craft in wartime were compared with the gunboats produced by our rival firms, the British Power Boat and the American Elco. Our task was the production of the heavier torpedo carrying craft essential to the attacking role. The gunboats on the other hand, using much the same size of hull with a form similar, if not exactly the same, had only to carry a gun armament, which allowed a faster and more manoeuvrable boat. If the roles had been reversed and the Vosper craft had carried only a gun armament their performance would have been outstandingly superior.

It was not until 1943 that the Admiralty decided, quite wisely, to combine gun armament with torpedo armament in boats designed and built by both Vosper and the British Power Boat Co. Then the real test arose and according to Sir Roy Fedden and his committee, some of whom came out in both types to have a run at sea, there was a marked superiority in both speed and manoeuvrability in favour of the Vosper craft, which for the first time were carrying the same load as their rivals.

On one occasion, when a new class of our boats was delivered to the Lowestoft base, the commanding officer of our latest flotilla asked for me to be sent there to hear at first hand of their complaints. These, strangely enough, consisted of design considerations relating to the bridge and operations room layout which had been arranged by a committee of operational commanding officers a year before. As can happen so easily, the operating officers, fighting in different waters from another base, frequently held entirely different views about bridge and plotting-room arrangements. Feeling unfairly treated on this occasion I pointed out that the latest comparable product from the yard of our rivals would not even answer the helm in the desired direction; which was true enough until a "fence" or horizontal fin was added to their rudders about half way down the rudder blade. This is a technique often to be seen on the wings of the modern aircraft, the function of which is to inhibit the spanwise flow which spoils the lift characteristics of the wing, which, for the purposes of this argument is represented by the rudder blade. It should, perhaps, be mentioned that the particular commanding officer wrote me a charming letter some weeks later when his boats had been caught out in a gale of wind on returning from a successful action on the other side of the North Sea.

There had also been an argument throughout the war, partly between George Selman and me, as to the optimum section for a rudder blade for these fast craft. Selman maintained that a modified aerofoil section represented the optimum, while I stoutly believed that the optimum was represented by the wedge section with a sharp leading edge and blunt trailing edge. The theory is not simple to express in non-scientific language. A rudder when running on a straight course is required to have a minimum of drag, while when turning it should produce the optimum lift coefficient; which are two antagonistic qualities. My view was that the typical aerofoil section envisaged, with maximum thickness at a quarter chord, beyond a certain speed of flow caused a breakaway or separation at an earlier stage in the speed scale than the wedge section; this caused a dead band or arc of rudder travel before it met solid water, and therefore started to develop lift or turning force. The accompanying sketches may go some way to explaining this.

Selman and I were by this time good friends, and set up a test with our own interests in a flume flow at the National Physical Laboratory at Teddington; and I believe I convinced him of my theory. Certainly it showed up very clearly in favour of the wedge section beyond a certain critical speed.

The experiment described above was not a perfect representation of the realistic conditions under which a rudder has to operate. Amongst other points of difference, there was no cavitation plate or bottom of the boat above the rudder reproduced in the experiment; nor was there the distorted slipstream coming aft from a high speed propeller, generating a spiral tip vortex. The realistic conditions under which a rudder operates in a fast craft should take into account, as well as separation effects, cavitation and air drawing from the surface, which is often called cavitation. The exact proportion in which these various phenomena affect the overall performance of a rudder is one of the many imponderables we have still to solve; and this

Fig 44 Boundary layer separation

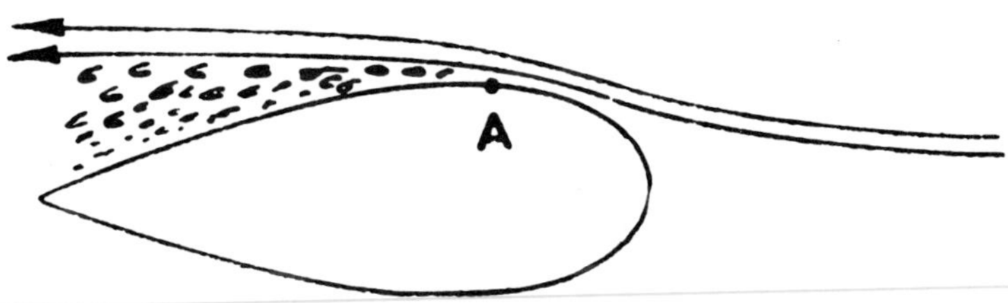

(a):—Separation phenomena

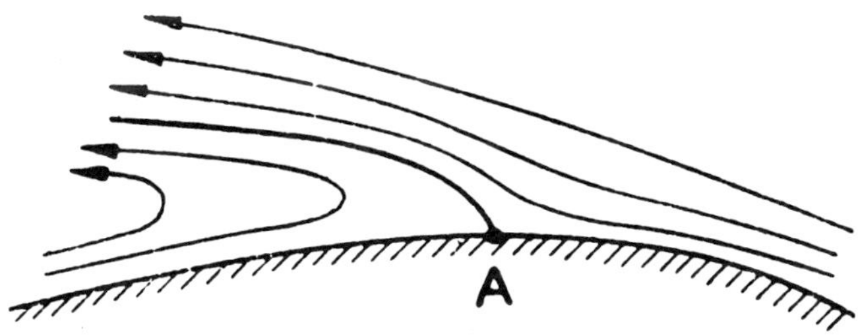

(b):—Streamline flow

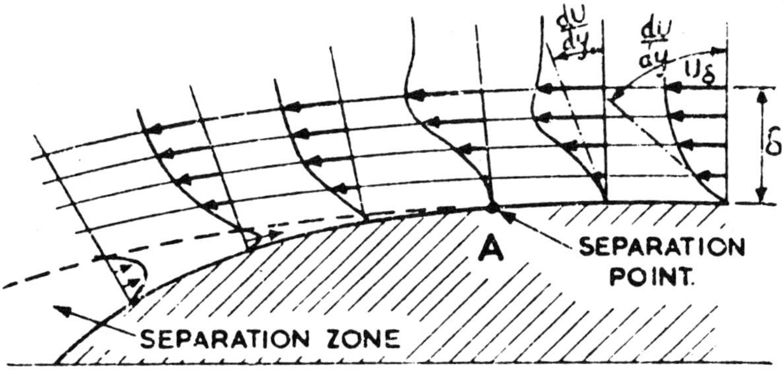

(c):—Velocity changes in boundary layer leading to separation

problem does not apply only to the fast patrol boat types.

However, on the whole George Selman and I were in agreement on most design features of the wartime boats, and the pity of it all is, looking back 30 years, that the Admiralty could not have put us together to work for the common good. It seemed at the time, and I fear now later that one cannot escape the same conclusion, that there was a large element of "divide and rule" in the behaviour of certain civilian sections at the Admiralty, which did little good to the morale of the fighting men.

Before the war the Admiralty could not have done what they are doing today. Now what must be considered a token order for boats and machinery is placed from time to time with specialised firms, including a contract to carry out a design study as a preliminary to completing the design and building of a few prototypes.

In the second world war no suitable engine was available at the outset of hostilities despite the fact Vosper had built a prototype at their own expense and had paid £15,000 of their own money to make available the detailed drawings and technique required for these Isotta Fraschini engines to be built in the U.K. All this, however, was to no avail. The Admiralty would not make up their minds that any staff requirements existed for such craft as MTBs until the fall of France, when the opposite side of the English Channel was occupied and much of the North Sea could not be used by any warships but the Coastal Forces. As it turned out we were, in fact, clearing prototype designs of hull and machinery until at least the beginning of 1944, when the war was three-quarters over.

One of the outstanding lessons of wartime coastal forces was that petrol machinery could not be used again and that an alternative must be found. The Fedden Committee, with understandable foresight, recommended the design and production of a light-weight high-powered diesel engine and sufficient encouragement was given to the Napier Co. to develop the Deltic marine engine capable in the first instance of a maximum power of 2,500 BHP. As I mentioned earlier, this committee could not find it possible to recommend the use of the rapidly growing number of gas turbine projects then on the drawing board, if not in actual production at that stage.

However, a certain amount of development and experimental work was carried out soon after the war, firstly with the mixed machinery in MGB-2001, which had a gas turbine of the Metrovick G2 type on the centre shaft, with two of the original Packards on the wings. This experiment, which was carried out on the initiative of the Admiralty and run by the Coastal Force Base at Gosport, HMS *Hornet*, was sufficiently promising for the Admiralty to design the two *Bold* class FPBs, *Bold Pathfinder* and *Bold Pioneer*. These two designs, one round form and one hard chine, were put out to tender and after consideration Vosper were awarded the *Bold Pathfinder* which was round form, and J. Samuel White was awarded the hard chine planing hulled *Bold Pioneer*.

These craft, 122 feet long, were powered with two Metrovick gas turbines each and in the first instance two Mercedes-Benz diesels, followed by two

Napier Deltics when these were ready. It was a delay in submitting our tender for these craft which saved my life, as I was booked to take a flight to the Bahamas in the *Star Tiger*, a Tudor aircraft which was lost with all hands. The delay, resulting from an appeal to the Admiralty to give us ten days more in which to prepare our tender, caused me to cancel my booking on this Tudor flight. Another of these aircraft was lost without trace shortly after this accident.

The resulting *Bolds* were both good sea boats but both were a disappointment so far as speed was concerned. Many important lessons were, however, learned, one of the most important being the devastating effect of sea water or salt spray drawn into the intakes in rough weather.

At the outbreak of the Korean war a programme of fast patrol boats, amongst other small craft, was put in hand with the utmost urgency. Vosper were allocated a type designed by the Admiralty not much different from those designed by us for the 1944 class of MTBs, with three Packard engines and the same gearbox arrangement. They were known as the *Gay* class and were reasonably successful once the trim had been corrected by the addition of transom flaps. The other class of FPB put in hand at this time was the *Dark* class, also designed by the Admiralty, and fitted with two Napier Deltics in a 68 foot hull. These craft were allocated to Saunders Roe of Beaumaris, Anglesea, who were to be the parent firm largely on account of a lithographic technique adapted from the aircraft industry which, it was claimed, would make series production easier and more accurate when other firms in the boatbuilding industry were ordered to build to this design. Whether this was so remained unclear, but the issue was confused by the extraordinary long time taken to clear prototype troubles in the initial stages and before the firms ordered to build these craft could get on with the job of production. The speed, once again, was rather disappointing as it was found most difficult to achieve much more than 40 knots, whereas 50 knots had been freely mentioned in connection with these craft.

By the time these two programmes had been completed the Korean war was over, so that there was time to look round and make a serious study of what could, and should, be done in the case of the next programme, the staff requirements of which we could guess.

11

The Braves and the Ferocities

Though not given this name at the commencement of studies, for the next generation of Fast Patrol Boats, *Brave* became their eventual name; though in the event only two were built for the British Royal Navy, they have been running for more than ten years. There must be few people in the world in any way connected with the sea who have not heard of them, even if they have not seen them.

It would be immodest for Vosper to claim all the credit for the obvious success of these boats, but it can be said that without the existence of such a firm, which had retained their specialisation and experienced design staffs and operatives, nothing approaching the success could have been achieved.

The British Admiralty in their wisdom, after two or three near misses, probably realised there was something to be gained by making use of the specialised knowledge in Vosper. Accordingly, when the staff requirements were approved, Vosper was awarded a contract to investigate and produce a design study. In the first instance it was more of a feasibility study, though I do not recollect that it was called by this name. The requirement, in short, was to produce a Fast Patrol Boat design capable of carrying a new weapon which was being developed, called CFS2. At the same time there was a firm requirement that 50 knots was to be achieved, or exceeded, at full armament load with half fuel aboard.

At this time it is not, perhaps, of sufficient interest to enumerate all the problems with which we were faced, or to present an account of all the meetings which took place between the Admiralty and ourselves. Suffice it to say that an official report was prepared after the whole of the design problems and prototype trials had been completed. This design study report and many details of the official trials were ordered and paid for by the Admiralty; it is an official document which can be obtained only by permission of the Minister of Defence (Navy), generally known these days as M.O.D.(N).

One or two of the outstanding features could be mentioned: the hull

Fig 45 The *Brave* class Fast Patrol Boat

form suggested by Vosper and approved by the Director of Naval Construction was that originally designed as a result of my request that we should produce a form in MTB1601 which incorporated the lessons learned during wartime operations.

After very careful consideration we came to the conclusion that the only machinery likely to come anywhere near to providing the power necessary for 50 knots at a practical weight and size would consist of three Bristol Proteus propeller gas turbines in conjunction with a combined reduction gear, clutch, reverse gear and vee drive. There was an alternative put forward to this arrangement, which consisted of a combined reduction gear and vee drive for each gas turbine with a smaller auxiliary diesel engine with marine reverse gear and clutch included, driving through an idler pinion into the main gear box. In this way the necessity to design a reverse gear capable of taking the full torque was eliminated.

After powerful arguments from a diesel engine manufacturer to the effect that we did not know what we were doing, and that a compounded engine put forward by them would have been the correct solution, we won our point and the Admiralty decided upon the installation incorporating Bristol (now Rolls Royce) gas turbines suitably modified for marine service. Then we prepared the detail design ready for building. Here it must be said that the structural scheme and stressing generally was decided upon to a large extent by the Admiralty of the time in conjunction with the NCRE (Naval Construction Research Establishment).

After some consideration the order was placed with Vosper for two of these craft to be built, the idea being at that stage that four more would be built in the following year.

As will be seen from the illustration (Fig. 45) the result was a particularly fine looking craft, and I had the pleasure of doing the initial running and testing of the prototype boat at sea and over the measured mile. In both respects she came well up to expectations and exceeded them by a handsome margin. Of course we had delays due to the sort of troubles which are apt to arise when prototype vessels of high performance are being tested, but I well remember a day when it was arranged that no less a party than the Commander in Chief, Portsmouth, the Director of Naval Construction, the Engineer in Chief and the Commander in Chief (Air) came out to see what we had produced and whether it was likely to be of value to the Royal Navy.

It turned out to be a good day for a demonstration with a wind blowing about force 6 from the south. This would be marginal, in my experience from the point of view of being able to run at or near full speed; so about an hour before the big shots were due to arrive I took the *Brave Borderer* out beyond the Nab Tower with our own crew, and found that by watching the sea carefully and making use of the transom flaps where necessary, I was able to give a demonstration at full speed in the open water much to the amazement of all concerned. It could not have been done at night because to achieve this sort of performance, in the sea running, required considerable concentration on approaching wave formations, and a highly responsive steering gear, which I am glad to say we had. Perhaps it would not be too immodest to say that considerable experience in handling this sort of craft in a seaway had a certain amount to do with the sort of demonstration we were able to give the big shots, who certainly went away duly impressed.

There was a sequel to this performance which, though unfortunate, was perhaps understandable.

After handing over, the first of the *Braves* was making a fast passage in the hands of a young man who had been aboard during my demonstration, and in trying to emulate the performance achieved in a seaway managed to damage the fore part of the structure, luckily not beyond the forepeak bulkhead. As a result considerable repair and reconstruction forward became necessary, after the boat had been towed by a salvage tug into den Helder, the Dutch Naval Base with good repair facilities. In the defence of the young officer concerned, who I am glad to say did not get into any serious trouble, I should add that the structure in way of the forepeak bulkhead left something to be desired. In the constructional design, which had been worked out between Vosper and the Admiralty, we had, rather unwisely, terminated the longitudinal or stringer members at this bulkhead, leaving a very severe hard spot or stress raiser, there not being even a rider plate between the bulkhead and the shell or planking. The structure forward of this point suddenly lost longitudinal support, except for the shell and a few frames, which latter were not of much use owing to the severe bevel at this forward part, where the angle between shell and transverse frames becomes excessive. This was subsequently considerably modified and no further trouble has been heard of in this region.

Since the first two *Brave* class of Fast Patrol Boats were produced many of this class have been built for the Navies of other countries and the speed realised on trial has reached as much as 57 knots on more than one occasion. The first foreign navy to order two boats, one of which was of different dimensions from ours, was the Federal German Navy and shortly afterwards the Royal Danish Navy. Then the Malaysians bought four of the *Brave* class, which presented rather a stiff problem as we had to guarantee 50 knots in trial condition in Malaysia. This was not too easy to achieve as the high temperature of the tropics involves considerable loss of power in the turbines.

However, by allowing for the fall off in power and watching the weights carefully we had sufficient margin in hand to achieve the necessary 50 knots. It involved fitting two sets of propellers. One set, with more pitch than the other, was used for running trials in England, where the turbines were able to give 4,250 HP, which resulted in some of the boats doing as much as 57 knots over the measured mile. As the power delivered was some 10 per cent less in Malaysia a set of propellers with less pitch was then fitted.

We have since received orders from Libya for much the same class of boat, fitted with guided missiles of a French make.

Two rather important points should be mentioned briefly at this stage as they affected our future production of the *Brave* class of boats sold to other navies besides the British Royal Navy.

The hull was constructed of wood making use of all the latest plywood and glue techniques initiated by our experiments in the construction of MTB1601. Also, instead of incorporating the reverse gear in the main Allen gear box as was done in the original *Braves* we used a pair of General Motors diesel engines fitted with their own clutch and reverse gear. These engines could be clutched in to the main reduction and vee drive box designed and built by Vosper so that slow running and manoeuvring were carried out entirely by the auxiliary GM diesels. This proved to be a very satisfactory technique and we have used this arrangement, as well as the latest wood and glue construction ever since. It involved what is called an S.S.S. clutch (Synchro Self Shifting clutch produced by S.S.S. Gears Ltd. of Isleworth) which protects the main turbine from being driven in the reverse direction which it is not designed to do owing to a unidirectional lub. oil pump.

Although we, and our customers, were very satisfied with the *Brave* class it has to be admitted they were expensive vessels having three Proteus gas turbines and their gearbox and propulsion arrangements. It was thought desirable to see if we could not design a craft to approach the performance of the *Brave* but making use of two Proteus gas turbines instead of three.

After some design consideration we produced a two engined craft, smaller but with a hull form likely to prove better in a seaway and still able to achieve 50 knots, with an armament almost, though not quite, as potent as that of the *Braves*. This Fast Patrol Boat was built at our own expense—in other words it was a private venture, used mainly for demonstration purposes, in order to persuade potential customers that what we were trying to sell them was indeed a potential weapon worthy of consideration. This is far more satisfactory for them than is the study of a brochure and drawings.

Fig 46 The fast patrol boat *Pfeil* built for the Federal German Navy

Most of the navies who sent delegates to us, and who were taken out to sea in the *Ferocity*, chose the three engined *Brave* class after careful consideration. This does not by any means detract from the useful function fulfilled by the *Ferocity*; and it is appropriate that the great help of Bristol, now Rolls Royce, should be acknowledged, as they loaned us the main machinery without which we could never have been able to afford the *Ferocity*.

As a fast craft, she was a good seaboat, but at first she could play some nasty tricks in certain conditions and though I, myself, could manage to feel the onset of this behaviour on the control and take the necessary action in time, it is quite certain that in the hands of the less experienced it could have become dangerous.

This has been described in detail later, where I have discussed the behaviour of these craft generally and in particular the Malaysian FPBs. Suffice it to say that by considerable experiment and much development running we cured the trouble to the extent that it was never a worry after about the first six months of her long and valuable existence.

One craft to the general specifications of the *Ferocity* was bought by the Federal German Navy; she was very successful, proved very good in a head sea, and had no nasty tricks. This craft, called *Pfeil* by the Germans, was sold to the Greek Navy after even more modification to prevent possible bad behaviour and she is giving great satisfaction as a sea boat. (Fig. 46)

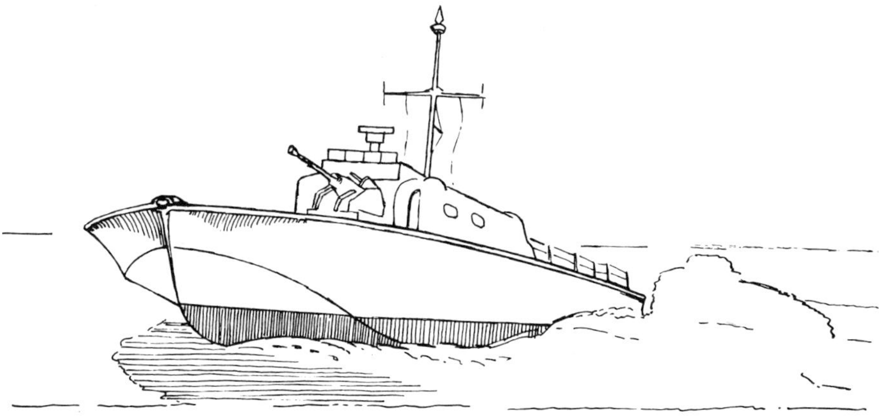

12

Some More Yachting

After the second world war we did very little sailing, but in 1946 we produced the *Viking* class of motor yacht in which we paid a visit to Deauville with Jeffrey Quill and his wife and my two daughters. Foreign currency was limited and all of us except the 14 year old daughter, Margaret Anne, had spent their allowance already. So we had to rely on her £45. Greatly to her indignation I asked Jeffrey to go with her when we reached Deauville to cash the travellers cheques into French money to avoid the distinct possibility of her spending it all in the shops on the way back to the ship. As it was we could only live for two days on this grand sum.

The *Viking* was a good type of little motor yacht 60 foot long with two Perkins engines; she cruised at about 9 knots with a possible 11 knots maximum. Deauville yacht basin at this time was fully tidal, having had the gates destroyed by the Germans on evacuation before the advancing Allied armies. It was, therefore, not too comfortable for a small yacht which had to take the ground at low water. Nevertheless the food was as good as ever in the Chalet Normande and the Regence in Trouville. Whatever else had changed as a result of the war lobster was still the same animal, as were the langoustines. The *Viking* with 12 others of the class were sold by the following year.

It was about this time that somewhat to my surprise I was elected to the Royal Yacht Squadron, a great honour, more particularly as I was chiefly a motor yacht man and also "In the trade"! I feel, in fact I know, Loel Guinness had a large hand in my election.

Our next venture was to convert a Fairmile "B" hull and install two of the excellent 6–71 GM diesel engines and a diesel generator. This became my own yacht *Sea Victory*, mentioned earlier. Loel Guinness, when he saw her, asked me to sell him another of these hulls and engines, and proceeded to convert her to a yacht much more thoroughly and luxuriously than we had done with our *Sea Victory*. This yacht became the *Sea Huntress*, which Loel and I with Teddy Phillips and Rupert Bellville took out to Gibraltar in

February 1949. We were held up by gales in Portsmouth before starting and two of the crew, including the chef, deserted us. Loel asked me what to do about it and I said the only possible thing to do before we lost any more was to take a deep breath, harden our hearts, and sail, despite the gale. This we did and got a fairly good hammering. After dark the lights started to go on and off, so I feared the worst and sure enough Bill, the mate, told me the batteries had broken their connection and were taking charge in the aft peak. This was a piece of damned bad work on our part and there was nothing I could do but struggle with these lead acid cells in an atmosphere of strong sulphuric acid fumes until I had stowed them with ropes and fenders to stop further movement. Bill helped me with his strength, and after struggling in this atmosphere for about an hour amongst lead acid cells on the move, we restored the situation, but not before I had ruptured myself and had been violently sick for the first time since the war. I had to have an operation on returning home and charged it to the firm—justifiably I think! Luckily it was not worse.

We reached Cherbourg in the early hours of the morning and moved up to the yacht basin later. After two days waiting for the weather to moderate we sailed for Brest; then came the long hop across the Bay of Biscay to Corunna. In some extraordinary way we managed to fix our position by sun sights nearly 50 miles to the east and made a landfall at Gijon! I cannot absolve myself from blame because I had taken some sights during the day before, but had never used the modern type of calculation as set out in the very helpful book by Mary Blewitt. My sight taking had been only very spasmodic since days in the Navy as Midshipman and Sub-Lieutenant. Then we used the long way of calculating the position lines using haversines. Suffice it to say that when we came out through the Bay the next year (1950), also in February, we managed so well that with Corunna and Finisterre in thick fog, we passed round them keeping well out to sea, and used our Bendix aircraft type radio compass on the Finisterre beacon, taking a 4-point bearing for distance. This was used during the war in bombers and other aircraft for homing on a radio beacon. The advantage over other types, where one has to get a "null", is that there is a pointer giving the bearing of the beacon continuously, provided the frequency is correctly set on. This time we called in at Vigo for a night and then at Leixoes, which is the port for Oporto. Here the motor of the 13 foot jolly boat broke down ashore, and despite all the efforts of Loel and myself to get it going in a rain storm, nothing happened until we found a mechanic from a nearby garage who looked at it, touched something, turned something, and off it went, much to Gloria Guinness' delight; who was able to call her husband and his guest from Vosper, clumsy muddlers.

After sailing from Leixoes we called in at Lisbon for a day then on round Cape St. Vincent to Cadiz. It was pretty rough from Cape St. Vincent to Cadiz and in order to do any reasonable cooking I had to hold Gloria steady while she cooked us some delicious omelettes. It was before we had fitted the roll damping fins, I hasten to add! This time, instead of leaving the ship at Gibraltar I went all round as far as St. Tropez where we got held up by mistral winds of gale force.

Others were held up in this rather pleasant port (in the winter) including Mrs. Reggie Fellowes, who was living aboard her lovely old time Camper & Nicholson steam yacht with clipper bow and schooner rig. I always remember a rather apt description this most amusing lady had of her chief steward who dropped a mince pie down the back of one of the ladies dresses. She asked the lady in question to excuse him because he was what she called "entre deux vins".

The following year Loel was on the look out for a larger yacht and wanted me to try to find him a suitable ex-naval craft. This turned out to be an ex-minesweeper built of wood under a wartime programme on the Great Lakes, which had adjacent forests. The type, about 150 foot long, was called BYMS and I well remember the miserable effect she made on us lying uncared for in St. Katherine's Dock near the Tower of London. We had her surveyed and though she was mainly constructed of one of the pines she was in good order and needed little done to the hull. So she came round to Vosper at Portsmouth where we set about carrying out a pretty big conversion, ready for her to go out to the Mediterranean where Loel had a small shipyard at Antibes, which was going to do the majority of the joinery and internal decoration.

This was a much more formidable operation than the previous *Sea Huntress* conversion and the 500 HP GM conversions were an unknown quantity. I probably took the wrong decision in getting the firm recommended by the English General Motors representative to do the job, as our people at Vospers wanted to have a crack at it. I was afraid that whoever did the job the engines would give trouble because they had endured a lot

Fig 47 MY *Calisto*

of wartime running, and I had enough to do without asking for trouble. As it turned out, the following year those engines were changed for Deutz—new diesels of more horse power—and very good they were.

I took the *Calisto* out for a trial trip after all our work had been done and all seemed to go very well except that in coming into the camber I rang down for an engine order which for some reason could not be obeyed, so we banged the stem into the wall of HMS *Vernon*. This required a graving piece to be inserted and we were then more or less ready to sail for the Mediterranean; except that the lady had a strong and justifiable complaint that steam was connected to the bidet instead of warm water. This was soon rectified and I never found out how this had happened because the only boiler on board circulated the radiators and domestic hot water.

Once again a delay of several days occurred before the weather was good enough to sail for Cherbourg, where we were to pick up Dolores Furstenburg, Gloria Guinness's daughter, by a previous marriage. However, we eventually got away all right and on arrival at Cherbourg we again had gales and rain, which was maddening; but to be fair it was February and one expects some bad weather then. Loel had very kindly invited my daughter Diana to come with us to keep Dolores company.

The next leg of the passage was from Cherbourg to Brest where we had to put in because one engine had failed for some reason that was not desperately serious. The chief engineer, Steve, who had served in the ship during the war, thought he could fix her up in about 24 hours. We had to enter the channel leading up to Brest in a thick mist and nearly collided with a fishing boat, which was on the wrong side of the channel. With only one engine running manoeuvrability was not all that could be desired.

After the engine was repaired and tested we were advised to obtain a met. report from the naval headquarters, where a commander on the staff told us we should have fairly strong south-west wind as far as 46°North, where calmer conditions would prevail. It blew all right down to 46°North, but after this it blew much stronger all the way to Corunna. I think they must have wanted to get rid of us at Brest as we were in the naval dockyard and flying the white ensign. Even French naval authorities do not always realise the Royal Yacht Squadron members also have the right to fly the white ensign provided they have a warrant—and anyway a lot of them do not like the Royal Navy. Even the British Consul at Malaga, an ex-Air Force officer as it turned out, came down to the harbour to inform us we had no right to fly the white ensign and was quite unpleasant about it till we showed him our Admiralty warrant.

Another alternative may be that you are mistaken for a naval patrol craft. Rather to our embarrassment the Trinity House Light Vessel dipped her blue ensign to us in the *Dimarcha* only the other day when we were coming out of the new harbour at Castletown Bere at the entrance to Bantry Bay in Ireland. Incidentally, anyone who goes into Bantry Bay or any of its ports or harbours must realise that since the big tankers have been going up to Whiddy's Island to discharge their oil practically all the buoys and lights marked on the Admiralty chart of Bantry Bay are incorrect, and have not yet been brought up to date. In particular, a submerged rock has been found near where the 350,000 ton tankers make their approach, so that this is well marked by newly laid light buoys. The only people who seem to have any idea of the situation are from the Irish Ports and Lights besides Gulf Oil Co. who operate and maintain all navigational facilities in the Bantry-Whiddy's island area.

This is a digression from the delivery of the *Calisto*, which was uneventful so far as Tangier, if we discount a fire in the dining saloon owing to the excessive proximity of the oil fired galley stove to the dividing bulkhead. There was also a Levanter gale of increasing force as we approached the Gibraltar Straits, and I was so glad to see Cape Spartel that I gave Gloria a pound I had offered to the individual who first sighted it.

In the *Calisto* we had two engineers, one of them an electrician lent by Vosper, who is now a chief electrical designer; thus at night there was always one man in the engine room. This was reasonable. When we had come out the year before in the *Sea Huntress* there had been only one engineer; to allow him to get some sleep Loel and I had taken the watch in turns on the bridge in the night and so let Steve go to bed. Thus each in turn was the only individual awake in the whole ship, and went down to the engine room once every half hour to see there was no leak and that the gauges read correctly. This was all right as the auto-pilot kept the ship on its course, but I am not very keen on leaving only one man awake in a yacht where there are a fair amount of other ships about. Also there is the possibility that he could fall and knock himself out or pass out for some reason, when there would be no one left in charge of the ship. Equally I do not like being the only navigator on board the *Dimarcha* as I am some age and might knock myself out or have some sort of seizure, when there would be no one left to make a precise and safe landfall.

Only the other day, for example, we sailed from Glandore direct for the Longships near Lands End, and this is a longish passage. When we reached the vicinity of Newlyn, our destination, I am sure one person in charge of handling the ship and navigation would not have been enough. Newlyn is quite a tricky little harbour to find, and to make it more difficult its white flashing light disappears as you approach on the correct course, leaving visible only the red light to starboard. All the local fishermen have complained of this. Of course in such conditions radar is of utmost help because one has to approach almost all the way to Penzance before turning up to the entrance channel for Newlyn. My navigator on this occasion was the famous Peter Twiss, whose charming and attractive wife Heather spent the whole passage reading "Airport".

From Tangier I had to go home taking with me Dolores Furstenberg, who had to return to school. She was a well grown 14 and very pretty. For a long time the French immigration officials at Marseilles would not let us in as they thought I was just a dirty old man leading a young girl astray; but luckily Dolores had her mother's passport with her picture on it and a note to explain why Dolores had not a passport of her own.

I had many happy cruises in the *Calisto* later and Loel was our inspiration for starting the roll damping fins or stabilisers (see appendix). He had crossed the channel on a rough day in one of the British Rail steamers and was so impressed that he said he wanted me to develop a set of fins for the *Calisto* which did not have to be retracted, there being no room for this. As told in the appendix, we first developed them to the point where we thought we could sell them in the *Sea Victory*. Then when we thought we were ready we told Loel, and he allowed us to fit a set at his yard in Antibes. At that time I was using the best hydraulic and control system that I could obtain for the job; this was undoubtedly the Fairey actuator and hydro-electric valve. At times since I have wished we had stayed with this very fine system, but as it was developed mainly for aircraft controls it was expensive and we had to try to remain commercial.

Loel now has an even bigger yacht built some years ago at the Bath Iron Works in U.S.A. and having the reputation of being a grand sea boat. She served as a patrol ship on the Atlantic seaboard throughout the second world war and came through unscathed. It was typical of Loel that having done so much work for him in the past and having accompanied him on so many cruises he invited me to do the first cruise with him in the *Sarina*; he also gave Vosper a lot of electrical work to do, besides fitting our roll damping gear with our own electronic control.

The *Sarina* is the most comfortable yacht I have ever lived aboard, and that even includes Arie Onassis' *Christina*; though I must say the *Christina* had a swimming pool on the quarter deck which, so far as I know to date, the *Sarina* has not.

Fig 48 MY *Mercury* 96-ft powered with gas turbines for Mr. Stavros Niarchos

What I missed, and I think Loel did also, was the fun of handling the ship and keeping a watch which we always did in the *Sea Huntress* and *Calisto*. With the *Sarina* there is a brass bound captain, mate and ship's company, which is not quite the same thing, as Loel will be the first to admit. Once this happens the ship is not your own. The same thing occurred eventually in the dear old *Sea Victory*. There was rather a large crew and they expected their meals regularly at the right time. The skipper was tactfulness itself and occasionally would ask one's advice on a navigational problem; but one skipper we had, who was ex-naval, was so good in his own opinion that one only had to ask him to do something for him to say this was quite impossible and he would want to go off somewhere else, or alter course in another direction.

CONTINENTAL TYPE FUNNEL

PDC TYPE FUNNEL.

TRAWLER TYPE FUNNEL

13

Offshore and Circuit Power Boat Racing

In 1961 the *Daily Express* organised, for the first time, a power boat race from Cowes to Torquay which became known as the *Daily Express* International Offshore Power Boat Race. The course was not straight down to Torquay, although that would have constituted an offshore race in itself, but took the racing boats well inshore for at least a part of the way so that spectators on shore would be able to see the competitors. The leading spirit behind all this was Sir Max Aitken, chairman of the *Daily Express*, a keen and expert yachtsman both in power and sail. The first race was in very heavy weather and won by Tommy Sopwith.

It was not until the following Boat Show at Earls Court, where we were exhibiting a rather attractive private venture craft about 45 foot long called *Show Boat* (Fig. 49), that Dick Wilkins, a good friend from days of John Cobb at Loch Ness, came on board at about 4.30 p.m. after what may or may not have been a very good luncheon, and said to me: "Can you design and build a boat for this year's Offshore Power Boat Race to beat that chap

Fig 49 *Show Boat*

Sopwith". I was naturally intrigued by this idea and asked for a little time to think about it. I rang up the next day to make certain Dick was really serious; and finding he was, we got down to work as quickly as possible, as there was not much time to get a prototype racing boat ready for a longish series of trials.

The regulations at that time allowed the hull to be built of any shape, except for not having transverse steps, within maximum w.l. length 40 feet. The engines, petrol or diesel, had to be available for sale to the public and advertised as such. With our really very extensive experience of the seagoing type MTB our first thoughts were to follow the general features of these craft, but scaled down to meet the 40 foot water line regulations. Considerable testing in the Saunders Roe tank (as it was then called) at Cowes enabled us to arrive at a good hull form from the point of view of resistance (speed) and sea keeping ability. One particular feature we found to be very beneficial was to eliminate all concavity from the transverse sections of the hull under the chine. In other words all frames would be quite straight below the chine throughout. It was found that this feature of straight frames did a great deal to eliminate the typical slamming in waves from ahead or on the bow.

The hull was based on our usual slightly warped form with a fairly sharply rising chine after about half of the length from the stern. We tested a fair number of forms with much the same characteristics and even went so far as to calculate, on paper, the very substantial advantages which could accrue from a twin hull form or catamaran. From these calculations, which were clearly set out in an article I wrote for "*Engineering*" called "A High Speed Craft Built to Win a Race" at that time, there could be no doubt of these advantages; but we did not feel we had sufficient time to enter into the design, construction and other problems relating to such an experimental

Fig 50 The first *Tramontana*—winner of the 1962 *Daily Express* Offshore Powerboat race

possibility. Now, eight years later, I wish I had had the guts to make the necessary decision, but apart from other matters the structure alone would have presented quite a severe problem at that time.

The first engines we chose were two Packard marine engines such as had been used during the war in MTBs and similar craft. However, the committee ruled these out as not qualifying, so we eventually purchased two CRM engines from Italy, which were really based on the well-known Isotta Fraschini marine engines of pre-war days, giving about 1,150 BHP each.

We managed to get her designed and completed ready for running by about the end of June, and this gave us plenty of time to get all the "bugs" out of the design and installation. We were even able to take her for a run down the Channel to St. Mawes, where the generous and enthusiastic owner was staying. Her name was *Tramontana*.

By the time the race was due to be run the boat and crew, under the command of Jeffrey Quill the famous test pilot, were absolutely on the top line and we had great hopes; though there was known to be some tough opposition from home and abroad. The Italian, Levi had designed *Ultima Dea* to be driven and owned by Gianni Agnelli, also much fancied was the American Sam Griffiths in the Ray Hunt designed *Blue Moppie*.

At the start *Tramontana* soon took the lead, but owing to some misbehaviour, possibly due to many causes which we sorted out afterwards, was seen to be lying third when the competitors came past the Royal Yacht Squadron Castle, from where Dick Wilkins and I were watching. We then drove over to Bembridge airfield as quickly as possible, where an aircraft was waiting for us, and on getting into the air we saw that just past St. Catherines *Tramontana* was holding a good lead with *Blue Moppie* running in her "slick"; but this lead the *Tramontana* never lost, and as might have been expected she really came into her own when faced with the open sea crossing in the pretty dirty weather of Lyme Bay—a tough passage of about 40 miles well away from land and shelter. When we arrived at Torquay, having landed at Plymouth airport, she was already in the winner's berth, having arrived there some 20 minutes before anyone else.

Admittedly she was the most powerful boat in the race, but this alone does not, of necessity, make for a winner, because proportionately more fuel load has to be carried to feed the more powerful engines, and this has to be done in a hull limited in size and so carrying proportionately a greater load, necessitating a structural scheme able to stand up to the rougher usage.

Opinions differ but most will say the race in 1962 was run in quieter weather than in 1961 when the first race was run. However, a close study of the meteorological reports and other weather data available does not altogether confirm this. It seems probable that at and just about an hour after the start the wind was stronger, but that by the time the real testing ground of Lyme Bay was reached the wind, though still strong, was a little bit more offshore than in the 1962 race. Anyway this did not in any way detract from the performance of the boat and those who manned her, who drove a magnificent race.

Fig 51 *Tramontana II*

After the race those responsible for the regulations altered the engine capacity permissible, which put the *Tramontana* out of court, so Dick Wilkins generously gave us another order to produce a hull capable of winning with machinery within the limit permitted. Dick was keen to use a British engine this time, so we studied the situation carefully and came to the conclusion that an installation consisting of four Jaguar racing engines would be within the regulations and still capable of driving a hull big enough to have the seakeeping ability of the first *Tramontana* at a speed sufficient to win if the weather was rough.

In a flat calm or near calm there would have been plenty of boats faster, but after all it is not unreasonable to suppose that in September there would be at least an even chance of rough weather on the day of the race. In fact, the strong probability would be that out of four races in four years three of these would have a good deal of rough weather for the competitors to negotiate. It is true, of course, that with the courses as they were planned there was a good deal of sheltered water in which the smaller, faster boats could gain a lead; but we still considered a boat capable of taking on the West Bay in its worst moods at reasonable speed would be likely to prove the winner.

We had learned quite a lot of lessons in 1962 with the *Tramontana* so that a certain amount of alterations were to be found in the *Tramontana II* hull form and structure. To have a chance of winning again we designed her to have much less weight in the hull yet enough strength to last at least one punishing race. In truth, we took a certain amount of risk with this hull in view of the fact that she would not, of necessity, be asked to accept the loadings inherent in running through a heavy sea more than once.

As it happens she very nearly did win, although the weather in the early part of the race was a bit light. However, as the Lyme Bay crossing was reached it was blowing up pretty well and there was a good rough sea running. Jeffrey, realising this was his chance, just locked the throttles full open on reaching Portland Bill and headed her for the Skerries. She was soon in the lead and going great guns when, unfortunately, a belt driving a dynamo and one of the circulating water pumps came adrift. This caused a delay of about 20 minutes, which just allowed Sonny Levi to slip in as a well deserved winner, while *Tramontana II* had to be content with third place. It looked as though she would win the All British prize, but later on word came that she had rounded a wrong buoy in the Needles channel area and so was disqualified.

This was really bad luck for owner, crew and all concerned with her. My sympathies were particularly with the navigator, Don Robertson, who had been so efficient that he had plotted all the courses at an early date when the buoy he passed was shown to be in the course and he did not realise the

Fig 52 Model being tested with and without appendages

(a) Craft fitted with Appendages

(b) Craft with Appendages removed Weight, L.C.G. and speed are in each case the same though (a) is much less resistful than (b)

course had been altered later. If this was not really bad luck I don't know what is?

Anyway the *Tramontana II* showed she was capable of winning given a bit of luck, and despite the light construction she distinguished herself in many following offshore races, though she never actually won more than the All British prize twice. So good was this modified hull in my opinion that I used it in my motor yacht *Dimarcha*, which has just completed 40 Channel crossings and four trips to Ireland and back. She will be described later.

The next time we had an opportunity to design and build a new offshore power boat racer was in 1966 or late 1965, when Dr. Savundra, well known at that time in the offshore racing world, asked us to design and build a potential winner for the 1966 *Daily Express* Cowes-Torquay race.

This time we did the most exhaustive set of tank tests at Cowes and came up with a hull form which seemed likely to be a world beater. At the suggestion of the tank superintendent, Bill Crago, we tested a series of hulls, all of whose performance was known in the racing world, against our hull with and without appendages fitted. The hull with appendages fitted, including propellers, proved surprisingly enough to be less resistful than when run without appendages. This can be seen quite clearly in Fig. 52. It is rather difficult to explain why this should be so, but it shows how important it is with these models of fast boats to make sure that they are tested with the appendages fitted and propellers working. Presumably the appendages give some form of lift, reducing the virtual displacement of the hull as a whole.

In this case the appendages, including propellers and shafts accounted for no less than 58 per cent of the whole resistance. It can well be understood, therefore, how interesting the so-called surface propeller can be if we develop it to the stage where no appendages, such as inclined shafting, propeller struts, and intermediate brackets, are located under the hull to generate drag. This surface propeller has only its lower half immersed, all the remaining shafting, etc. being above water. This will be discussed at greater length later in the chapter on the smaller circuit racing boats.

The *Flying Fish* as the boat was named (Fig. 53) was fast and a brilliant sea boat, as was shown in her one and only race, when she took the lead in the rough stuff. Then she sank for no known reason—but probably due to a propeller blade piercing the bottom.

Just a few thoughts on the offshore power boat racing hulls may be of interest here. There can be little doubt that any hull, to have a chance of winning, must be very good in a head sea. Amongst the early offshore racers, the Ray Hunt designed *Blue Moppie* was probably one of the outstanding boats. She had a deep vee form and was very beamy, relatively speaking. The form seemed to me a very good and soft rider, but when the waves from ahead became large and fearsome there was a tendency for the boat to be thrown into the air by one wave and then to land very awkwardly on the face of the next wave, causing a nasty slam and a severe shaking up through-

Fig 53 *Flying Fish*

out the hull. To some extent this tendency to fly off the top of an oncoming wave would be due to the relatively large beam forward combined with light overall weight. Our first *Tramontana* had a fairly large beam, but by virtue of the running trim, due to the warped sections approaching flat right aft, combined with relatively large weight, she did not have this tendency to fly into the air. The forefoot remained well down, so that the veed forward sections would enter the water at the right angle and avoid a slam except in the worst conditions.

At any rate the attempt was made consciously in the design to avoid bouncing and instead to ensure that the veed forefoot entered at the correct angle for optimum encounter. All successful offshore power boats in these days aim to "enter" the waves in this way, and it will be observed that the beam to length ratio is now much reduced in the more successful craft, a good example being *Cigarette*, Don Aronow's fabulous winner of the 1969 Cowes-Torquay-Cowes race. It is even possible to say, despite the wonderful record of these boats, that the beam has become rather too small, and it is rumoured that they can play some quite nasty tricks if the pilot is not careful. They can, under certain circumstances, "spin" due to the forward sections "digging in". Then the boat swings round very violently as the yawing axis momentarily moves so far forward, due to this "digging in" of the bow sections, with the fore and aft axis at least temporarily at an angle to the line of advance of the centre of gravity of the boat as a whole. One could, perhaps, call this a "broach", which is really another nautical term for the "spinning" characteristic. Such behaviour is quite well known amongst the smaller and relatively faster circuit racers, both inboard and outboard, and is really little different from the misbehaviour described in the *Ferocity* and certain of the *Braves* where air drawing was found to cause the undesirably low trim.

So far as my experience goes in these matters of designing the racing hull to slice through a wave, using its well veed fore body to throw the water encountered well away from the boat, it has always seemed to me that Sonny Levi, who has designed many very successful offshore power boats, had the first concrete ideas on this subject. In other words he realised that to go fast up wind it was necessary to slice through the oncoming waves rather than jump off them and come down with a terrific slam on the slope of the next wave ahead.

The one most important position not to get into when dealing with a heavy oncoming sea is to be thrown over so the boat lands on its side, which is flat and not usually so strong as the bottom sections. If a nasty wave appears on one bow and looks likely to cause the boat to be thrown on to her side the best technique is to steer towards the wave in the last few seconds available.

Sonny Levi's plan form has, for some years at least, been almost triangular, or Delta shaped, with the narrow sections forward, thus making it easier to slice through the oncoming waves. The shape has proved highly successful and many well known boats have been designed on these lines, such as *Surfury*, which has won the Cowes-Torquay race twice in the capable hands of the Gardner brothers. Many more could be listed, but one very successful boat which comes to mind is the *Fi-Fi*, a most successful circuit racer in the hands of the distinguished designer Don Shead. This configuration obviously involves the designer in keeping the centre of gravity well aft where the transverse sections are greatest.

Of course, it can be overdone, and there have been one or two cases where the penetration of the wave ahead became uncomfortable to the driver and his team. In one case also a hull on these lines took a dive in an awkward wave and actually hit the bottom of the Seine, which is not very deep where the famous Six Heures de Paris takes place. However, these are the exception and many good boats come to grief in one way or another in this race.

I cannot omit mentioning what to my mind is the greatest single advance made in the technique of designing and building these offshore boats. I refer to the Mercruiser outboard drive with prop. axis adjustable from inboard called Power Trim by the makers.

Making use of this ability to adjust the line of thrust of the propeller enables the optimum equilibrium to be achieved. In effect, the boat as a whole can be made to assume such a trim that almost no part of the hull is wetted, and this an absolute minimum right aft. This, in turn, leads to very

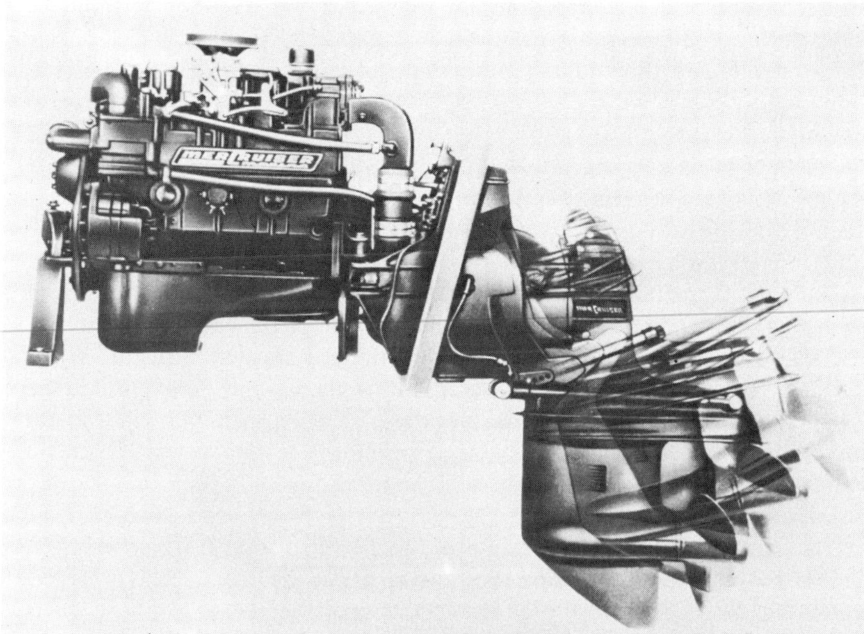

Fig 54 Mercruiser Power Trim

low resistance and highest speed. As Sonny Levi said to me one day recently, if you fitted this form of drive to a log of wood you would still have a winner! Of course, what I have been describing is the setting up of the propeller axis to give optimum trim for low resistance, but it can equally be the case that in a seaway the trim will require to be altered one way or the other to suit the conditions prevailing, and this can be done from the driver's control position—an immense advantage.

So many interesting problems arise in the design of these fascinating offshore power boat racers. To cut down appendage drag some boats drive the power of their engines into one gear box, from which the output to a single propeller runs along the centre line. Other designers avoid this because, while being undoubtedly more efficient, a severe problem is set by the torque effect from one propeller. There are, of course, ways to offset or eliminate this torque while running steadily in the water—for instance, a slight offset from mid line of the engines and shafting. If necessary other weights can be used to give a static list which is ironed out when the engines are running at full speed.

A little bias applied to the leading edge of the rudder can do quite a lot to help. When, however, the whole boat, propeller, and rudder included, leaves the water a magnified torque effect is applied to the boat when it reaches the water again and the propeller starts to "bite" solid water. Then a nasty list caused by torque is apt to throw the boat on to her side with all the unpleasantness this entails. It is for this reason that a great majority of designers will prefer to transmit the power of two engines through two opposite handed propellers. Although appendage drag is greater, being almost duplicated throughout, nevertheless the problem of torque effect is eliminated.

To run fast into a head sea a water ballast tank is frequently used, which is a great help in keeping the well veed forward sections well in the water. To be effective the means for filling and emptying these tanks must act quickly and the amount must be easily adjustable. Aft ballast tanks are also sometimes fitted.

The hull form itself can be argued about until the cows come home, whenever that is. There are still many in favour of the round form hull for real sea going, and some very fine performances have been put up by, for instance, *Horatia* designed by Peter Thornycroft and *Spirit of Ecstasy* designed by Arthur Hagg; but despite these excellent results I will still back the well designed planing boat to win in even the worst weather. They may be more uncomfortable but they will get there first just the same—at least that is my opinion.

One can argue about the design and location of the propeller bracket—some will opt for a bracket and toe bearing aft of the propeller, while others prefer the more orthodox placing of the propeller bracket ahead of the propeller. Whichever way is decided upon the utmost skill must be employed in designing this feature to offer the least possible drag while retaining the necessary strength.

The rudder(s) want careful consideration and there is a lot to be said for keeping them out of the slipstream—in fact I believe it to be an essential where the fast cavitating propellers are used. With a fully cavitating propeller, which really cannot be avoided in these fast high powered boats, the cavitation blister which spreads from the blade back will avoid collapsing on the propeller itself, but will collapse on the rudder blade if it is in the slipstream. This is where the damage is done, and one at least of the competitors in the 1969 *Daily Telegraph* Round Britain Race had the blade of his rudder sliced off each day and a new one had to be fitted, until the owner of *Maltese Magnum*, John Kennerly, eventually had to give up.

All this does not apply to the, so called, outboard drives, an example of which is the Mercruiser drive mentioned above, because the steering of the boat is carried out by a change in the line of thrust from propeller(s) so no rudder blade is required. The same, of course, applies to all outboard engines.

The exact position of the so called "spray strakes" can be argued about for a long time without coming to a satisfactory conclusion on paper or even on the model, but no one could begin to deny their effectiveness, especially when used with the deep veed type of hull, which would be inefficient were it not for the substantial strakes placed so as to reduce wetted surface to a minimum at high speed.

Then we have steps to argue about, as well as catamarans, three pointers, *et al.* There is really no clear cut finality as yet in designers' thinking, though

I have a hunch that catamarans and three pointers are shortly going to become increasingly popular in offshore racing. Of course cats and three pointers make use of aerodynamic principles to help them lower their virtual displacement and therefore to reduce their wetted surface and resistance, but aerodynamics affects everything that runs fast through the air, even if it is over the water, so why not make use of what we can if it helps?

While on the subject of offshore power boat racers no one can avoid mentioning the brilliant performance of all the Fairey Marine types, which do consistently well in the most capable hands of Peter Twiss and others. The latter showed his superb driving when a few years ago he drove Merrick Lewis' *Thunderbird* into second place. This boat had a fantastic career when it was subsequently driven by Tim Powell and Norman Barclay in the Round Great Britain race when, if they had not run aground in thick fog, they would have won the race. As it was they were second and a great performance was put up by the Fairey boats with Ford engines, who won the team prize.

Lastly I should say a few words about circuit racing; that is, the racing of small motor boats driven by outboard engines or inboard engines up to a maximum cubic capacity of 2 litres.

The outboard classes have tended to monopolise the scene in the last few years. Not only do they tend to win their own classes, which reach up to 1,600 cc, but in almost all recent classic races they have tended to win the overall prize as well. The reason for this is probably not difficult to understand when it is realised the enormous resources and potential market of the major builders such as OMC (Johnson and Evinrude) and Mercury—both built in the U.S.A. Both of these enormous organisations have competed with each other for many years and have very well financed competitions departments.

The inboards on the other hand have not really bothered so much and with the exception of Volvo and BMW, they seem to be quite satisfied with their automobile sales or marine sales for the cruising types of craft, in which the utmost in turning and power to weight ratio does not seem to matter to them.

The above remarks only apply to what are essentially circuit racers because in the larger classes, such as Class I which we have been discussing above, practically always an inboard engine will win. As a rule these are marinised versions of large U.S. automobile engines.

In what is now called Class III, with which a considerable number of offshore races are run, the winner will frequently be a boat installing two or more outboard engines. However, the inboards have more of a chance here.

In recent years the catamarans designed and built by Molinari have tended to win the classic circuit races such as the Paris Six Heures and the Berlin Six Hours. In all recent winners either a Mercury or Johnson have been installed; it is only recently that the Johnsons have seemed to come into their own with a near win in Paris. A Johnson engined cat was leading until the last few laps when hull damage put it out of the race. In Berlin,

Fig 55 Start of the Paris 6 Heures Race

two weeks later, a Johnson engined cat took the honours with a good margin to spare. However, Molinari had no need to be sorry because the winning Johnson outboard delivering about 140 BHP, was installed in a Molinari designed and built catamaran.

The charm of these races for many is the fact that they are run on rivers (the Seine in Paris and the Havel in Berlin) so for the spectators there are plenty of spills and thrills. Also, thanks to the relatively small size of these little racing boats, they are not too outrageously expensive—they are expensive enough, though the first class drivers and boats get quite a lot of help from the engine builders.

I have designed two or three fairly conventional monohulls, all of which have had inboard engines installed, and on the whole, considering the competition we are up against internationally, they have not done too badly. In *007* we have altered her several times until she is a delight to drive and can do about 60 m.p.h. with a BMW racing engine installed—more with the specially tuned injection engine. The other boat I designed was *Badger II* with a racing Volvo which did quite well for a time.

Now, however, I am designing on entirely new principles in hopes of at least competing with, if not beating, the cats—we shall see! Initial running has been quite promising. The *007* has been through various phases as her pictures will show (Figs. 57 and 58). She started life with a Volvo racing engine and was then considerably altered to take a BMW engine giving

Fig 56 Ram Wing racing boat *Advance*

Fig 57 007 as built originally

Fig 58 007 running in Rouen 24 hours race after alterations

somewhat more power.

The enormous amount of work and enthusiasm put in to all these boats cannot be described without mention of the sterling work of Nigel Tunnicliffe and Mike Trimming—as well as a lot of others including the crews.

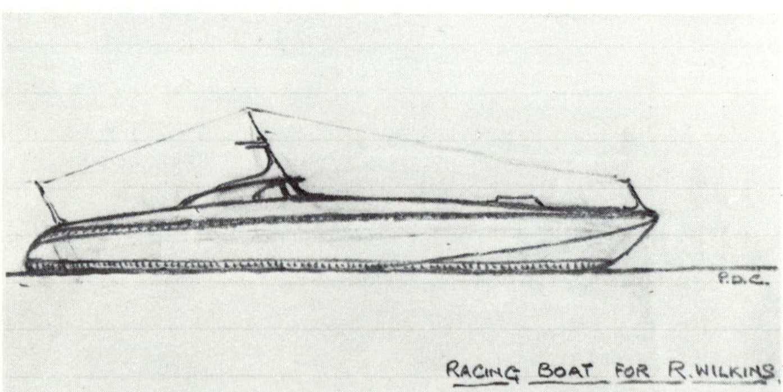

14

Dimarcha

The *Sea Victory* lasted us for many happy years until the new M.Y. *Dimarcha*, designed by myself for my own requirements, was handed over in 1965. A full description of the design and equipment follows in the form in which it was given to the press at her demonstration.

"Exactly when I decided to design and build a yacht for myself, or rather, ourselves, is not clear, but I do certainly remember getting out what I still think is rather a nice 80 footer, steel, round form and diesel driven. This would, in many ways, have been ideal for us but frankly, and although my firm Vosper did what they could to reduce the price, it was much too

Fig 59 MY *Dimarcha*

expensive for me to shoulder the burden.

The strong probability is that the germ of the idea was born a few years back when it became obvious that our very dear old friend the *Sea Victory*, an ex Fairmile 'B', was running out of useful life and that the time was approaching when I could not expect so much help from my firm in running the yacht and providing the crew.

Of course, not everyone wants, or thinks he wants, the same as us. Perhaps the majority would want a sailing cruiser with a motor to help out. To some extent this is a matter of age and in my case, also, it is what I have been used to running about in—although possibly on a rather smaller scale. Also, I fear, it has to be admitted I am a fidgety and somewhat unrelaxed character and like to get from A to B fast. Or put another way, it irks me to take ten or eleven hours in crossing, for instance, from Portsmouth to Le Havre or Deauville, frequently in some discomfort.

At the significant time, in formulating my ideas on this design, we at Vosper were running two relatively small fast cruisers, each of which had many advantageous features, including what I considered to be somewhat outstanding hull forms; both were well in the planing regime. These hulls were represented by the *Show Boat* (Fig. 49) designed and built by Vosper as a private venture and exhibited at Earls Court in January 1962. The other was the *Tramontana II*, an offshore power boat racer (Fig. 51). In neither case were these two craft outstanding as commercial or competitive successes, but, to those in the know, they had that certain something and in a number of respects were somewhat outstanding.

My yacht is designed around a really pleasant day saloon (Fig. 60) from which one can see out of the windows what is going on outside at surface level while remaining seated. The other, and obviously important, feature must be the provision of as near as possible 20 knots max., say 16–17 knots cruising with reasonable seakeeping qualities at speed.

To make any sense from a reliability and price point of view it was necessary to see what was available as a standardised article in the way of a diesel prime mover, which could provide the power required for reasonable cost and with satisfactory power to weight ratio. After very close consideration and consultation with a lot of my colleagues, as well as rivals, in the world of fast cruisers, I decided on the GM8V/71 offering a maximum shaft power of 320 at 2,300 revs/min. and with an overall weight 3,130 lbs. What was, perhaps, more interesting, the makers offered 290 BHP for continuous operation which is, in fact, a very good and useful figure indeed (Fig. 61).

Design of any vessel is a tentative process, but the form of *Tramontana II*, with only slight modification, at an overall length of 60 feet, was eventually decided upon. This was not, of necessity, a quite obvious selection because, in truth, for the speed and displacement predicted here we should almost certainly have used a round form from hydrodynamic considerations, but the planing form has certain advantageous features, not least of which is very adequate beam, which helps one to get in what is wanted in the way

Fig 60 Day Saloon in MY *Dimarcha*

Fig 61 Engine room of MY *Dimarcha*

of accommodation within the hull.

There were also strong protagonists for the *Show Boat* hull versus the *Tramontana II* type at this relatively low speed/length ratio and I could only settle this by running models of both of them in waves at the Westland Tank at Cowes at the appropriate speed. Even then I found it necessary to obtain quite unbiased opinions from persons unconnected with the project to watch the runs and make the decision for me. This opinion, incidentally, agreed with mine but it was by no means an immediately obvious one.

Beside the importance of the day saloon the decision was made to put all the machinery right aft and drive through vee drives which offers by far the best opportunity of avoiding the bugbear of the high speed diesel cruiser, which is noise. The great majority of fast yachts, in order to make use of the space available to best advantage, will put the machinery under the floor of the saloon amidships. The resulting layout makes excellent use of the space available, but it is very rare that you can hear yourself, or anyone else, speak in the vicinity of the machinery, which detracts very greatly from the pleasure.

As a consequence of this decision it has to be admitted the sleeping cabins are not so big as they might be and perhaps there could be more of them. However, this is a limitation we know about and accept, though many might not. It has also to be admitted we have not more than a couple of showers, whereas a bath will be shown in the layout drawings of some other craft of similar dimensions. This we do not consider realistic so far as baths are concerned on account of the severely limited amount of fresh water which can be carried in a yacht of this size. We have basins in each of the two forward guest cabins and the forecastle, while the two aft cabins share a really large folding basin between them in a central lavatory compartment.

Steering Gear

The wheelhouse has been planned as one of the principal features. Normally the directional control will be effected through an automatic pilot; the course required is set on to a heading indicator by a milled knob and the auto-pilot cum electro hydraulic steering do the rest. As well as the small steering wheel there is an override to this by making use of a small lever or "stick" on the steering panel when a form of bang bang or on-off control results. The actual steering assembly, in this case, is of a prototype nature and more complex and powerful than is really necessary for a yacht of this size, but Vosper are using this yacht as a trial horse for this steering system, which has been developed in combination with S. G. Brown, making use of their Arma Brown gyro compass as a datum in conjunction with their auto-pilot. Vosper are adapting their electro-hydraulic units which have been used very successfully for some years in their roll damping installations. The hydraulic spool valve is operated by an electric motor instructed electrically from forward by the auto-pilot steering wheel or override.

The connection between the control position forward and the rudder operating gear aft is by electric cable which, with suitable duplication and standby arrangements, eliminates the necessity for shafting, rods and levers, etc. etc. running through the ship. For steering the ship there is an alternative remote control on a flexible lead at an outside position on the upper bridge. There is also a simple tiller system aft for the ultimate emergency when even the battery operated standby pump and hydraulic cylinder have failed.

Radar

Decca type 202 is fitted with presentation in the wheelhouse adjacent to the steering console. This is unusual in such a small yacht and the revolving array involves quite a problem in mounting it high up on a reasonably attractive looking structure. In this case I have elected to use a goal post type of mast which has many practical advantages, although it has to be agreed that to some eyes, at least, it is aesthetically a bit unusual—or takes a bit of getting used to. The alternative would have been a mast somewhat of the *Mercury* type, but it was in fact difficult to locate amidships for a number of reasons.

General Appearance, Interior Design and Decor

Much more effort than is probably realised has been spent, from the early sketching days till the approach of completion, on getting the overall appearance and effect attractive and impressive, if not striking. This is by no means easy, but a yacht of this length is particularly difficult as it calls for full headroom in at least the wheelhouse above the main deck. Even if one manages to get an attractive overall appearance from the original sketches, which, of course, is the first move in the game, there are so many practical matters requiring compromise which crop up that a very conscious and almost constant effort is involved in retaining this attractiveness.

Apart from anything else, unless a perspective sketch is made (which is extremely difficult to do accurately for even the most experienced artist) the only way to know what a yacht will look like is to make really good models and it is important to have plenty of detail such as rails, windows, masts, etc. It is only in this way that any idea can be obtained of the realistic third dimensional appearance bearing in mind that the normal ship's drawings as prepared in project or ship office or as initial sketches, will be two dimensional; thus they represent only a unique and, for various reasons, somewhat compromised view of the ship.

Of course most of us concentrate very much on this beam on, or side elevation view and it is certainly a good start to get her right from this angle. However, it is an angle comparatively rarely seen in practice and the head on, bow on, and quarter views are equally important.

Appearance, and what is nowadays known as styling, is remarkably important and is very much a matter of personal taste. It takes very little to muck up the appearance generally, but great effort and much determination is required to retain an attractive line and style throughout. My first rather rough effort on a blackboard is shown at the end of this chapter.

So far as interior decoration is concerned we have done what we can to make it attractive and the joinery workmanship has been really beautiful. Here we selected a mahogany with a rather light reddish colour and for such items as drawers, cupboards, etc. the timber required has been cut from the same log and has been used throughout. This has been treated with Elastogeen Vernis Glycerophtalique to retain the colour without stain and by sealing the grain make it impervious to stain, which can be washed off as easily as, for instance, Formica.

In the saloon and cabins a slightly patterned Vinyl finish has been applied to the remaining surfaces. A feature which requires quite a lot of consideration in a layout such as this comprise the staircases, rails, banisters, etc. A happy mean has to be struck between what one would have at home and the minimum strictly functional hand rail of wood, plastic, metal or a combination, in a yacht such as this.

Wheelhouse and Control (Fig. 62)
This is one of the most important features of the yacht and has been planned to be functional and at the same time attractive. Here there is the control panel including engine and steering controls and instruments. There is the radar presentation, the gyro compass and the electrical switchboard, not to mention the *Hector* echo sounder and an aircraft type of magnetic compass overhead, also the radio ship to shore and radio D.F. beacon loop.

We have the galley and refrigerator on the port side. Now this may or may not be a success—plenty of people criticise me for it. On the other hand it was remarkably difficult to find anywhere else for it which did not seriously reduce the saloon or cabin space. Much depends upon how much one intends to use the yacht for providing main meals.

When actually at sea, under way, it is virtually impossible and not at all practical to provide anything in the way of set courses at a laid table. On the other hand in most cases a passage will be of relatively short duration with this speed, so that a little forethought in the way of preparing sandwiches or preparation of pies, soups, patés, etc. can keep the wolf from the door quite well until harbour is reached. At sea, the ability to have a jolly good cup of coffee or soup and to open a tin of sardines or similar, goes a long way to meet requirements and this is not difficult to organise. We have the immense advantage by virtue of having 240 A.C. supply of having a really rapid electric kettle and toaster.

In harbour things may be different, but in our own case we tend to go ashore for an evening meal, leaving the field clear for the ship's company. It has to be doubted whether the traditional roast and two veg is particularly easy or appropriate at sea in a small yacht like this, but suppose it is not by any means impossible, especially if the latest techniques are used and perhaps a pressure cooker.

We have a rather severe problem which may not yet have been solved. This arises from the fact that the control of the yacht and cooking take place in the wheelhouse, which has no direct access to the deck or forecastle and

Fig 62 Pilot house of MY *Dimarcha*

at present has no opening window. The latter, we hope, will be met with by installing very comprehensive demisters and also an outboard control position between the legs of the mast, adjacent to the steps and hatch provided at the aft end of the wheelhouse communicating with the open bridge and sun deck.

Direct communication with the forecastle from this bridge is rather complicated by the fact that we have to use steps and rails over each side of the deckhouse from the bridge position in order to get to the main deck and forecastle hatch.

The alternative, which one would like to be able to provide, would consist in the provision of opening side doors from the wheelhouse together with a bridge screen at each side. This is remarkably difficult here owing to the amount of important equipment it is required to locate inside including the chart table and drawers, also owing to the fact that we have located the galley on the port side. So far we have found no outstanding difficulty once one is used to it.

Hull Structure

This consists mainly in plywood panels glued and scarphed together to form the shell laid on plywood frames. A particular feature is the rounded gunwale which can be seen in the photo. This is made up of laminated timbers running fore and aft.

Electrical Installation

The basis here is two 5 KW Onan diesel generating plants to carry the normal 240 A.C. 50 cycles. These units start automatically when the load is sufficient to require them. There is a police light system supplied from a 24 volt battery system which obviates the necessity for running the generators during the silent hours. These circuits cover such things as bunk reading lights, sufficient lighting to enable the W.C. to be visited during the night without the whole ship being disturbed by the starting of the Onan and also the same applies to one or two reading lights in the saloon in case one of us does some late or early reading or sketching—the latter being somewhat of a habit of mine; in fact I do my best work about 5 a.m. at which time most of us are decidedly not at our best.

Incidentally the two main engines provide a 24 volt D.C. supply from generators running in parallel which, amongst other functions, operate the steering gear, and wheelhouse controls including the Arma Brown gyro compass through a battery of reasonable capacity.

Boats

Our only boat, apart from inflatable raft of the RFD type, is a Boston Whaler with Mercury 35 HP outboard. The Boston Whaler formed of fibreglass is a rather well thought out design by Ray Hunt, 13 feet long and strictly utilitarian. It probably needs a bit more of a shelter to make it really practical for our purposes, but this is not difficult. The advantage of the multi hulled or semi catamaran form is that it has immense stability and one can use it to great advantage, for instance, for water skiing or bathing parties. These boats are very fast and plane nicely and easily so long as there is not too much of a sea running, when they can begin to slam quite substantially, but all boats of this sort represent a compromise. Rightly or wrongly we have elected not to have electric starting owing to the weight and the hand starting does not seem too difficult.

Such running as has been done up to date shows her to be a great success.

We have found that, as always, the weight of the yacht as a whole has come out a little more than was estimated; therefore the speed is, so far, a little disappointing to the extent of up to a knot at 16½ cruising speed door-to-door; however by using the forward ballast tank an increase in speed, with reduction in slamming in head seas, is achieved. It is obvious we shall eventually gain by fitting electrically operated flaps at the transom, and this will probably be done during the winter months.

The yacht seems quite easy to handle from inside the wheelhouse in all conditions. She is a bit lighter on the water than some and catches the wind a bit unless one is careful, but the engines are very powerful and manoeuvrable through the Bloctube controls. Judicious use of the engine power makes her very satisfactory and easy to handle."

This yacht, after six years, has been an unqualified success apart from a few teething troubles, mainly with the hydrostart equipment in the first year. She has crossed the Channel 40 times since completion and has crossed from

Lands End to the South Coast of Ireland each way four times. She cruises at 15 knots, which almost halves the time taken by the *Sea Victory*, thanks to the additional power provided by two 330 HP General Motors 8/V71 engines. We have been to Paris and back six times, and separate cruises to Rouen and back five times; the main object of these trips was to observe or take part in the two annual classic circuit races, the Paris Six Heures and the 24 Heures de Rouen.

Most of our cruises have gone smoothly, but on the first occasion of crossing the Channel I managed to flatten out the 24 volt battery from which all navigational services, including steering, gyro compass, windscreen wiper, radio and D.F. loop are operated. We found ourselves—only three of us—myself and wife and Fiona Arran of motor boat racing fame—obliged to rely on the aircraft type of magnetic compass fitted in the wheelhouse, while for steering we could only use the engine throttles differentially. We arrived in the Seine estuary not much later than estimated but had to waste three hours before entering the tidal yacht basin at Honfleur. Imagine my annoyance—mainly with myself—when I found the lock gates closed and on checking found I had been using the tide tables of the previous year. This led us to make fast outside the basin along with all the fishermen, who do not mind going aground. We then went ashore to dinner at the Ferme St. Simeon, run by our very good friend Monsieur Goulet. On return we found the 220v. generator stopped as the yacht had gone aground and sucked up mud into the cooling system. Luckily there was an overheat safety device which had shut down the set before damaging the engine.

Indeed this was a day of accidents, but we had arrived in the right place and also learned a lot. The flattening of the 24v. battery was not altogether my fault. To put the main engine driven generators on circuit it was necessary to press two starting buttons on the dashboard, but no one had told me about this, nor was it mentioned in the operating instructions handbook. The tide table nonsense, as well as leaving the Onan 220v. generator running when we knew she would take the ground, was definitely my fault. By getting up at 4 a.m. I was able to pump up the hydraulic bottles to a satisfactory pressure for starting both main engines, so Victory and myself alone took her out of the basin at Honfleur and up the coast to Deauville where facilities were better. Luckily a kind gentleman on board a large sailing yacht was able to take our lines from Victory and allowed us to lay alongside of his craft. We remained in Deauville for at least three weeks, having various troubles corrected. Peter Rowland was with us as engineer for the return passage, which was a very great comfort.

On only one other occasion has it become necessary to return without the use of our electric hydraulic steering gear and auto-pilot. It was when, in the middle of the Channel, we struck something very hard, which jammed the rudder and caused us to return on engine throttles alone.

An extraordinary chain of misfortunes accompanied our return from the Paris Air Show in 1967. We made fast in the Bassin du Citadelles at Le Havre after running down the Seine and across the estuary. The weather was hot so, rather foolishly, we left the hatch to the wheelhouse open. It was not until I got up to listen to the weather report at 0640 on the B.B.C. Light programme that I realised we had experienced a burglarly. Our binoculars, ash trays, cushions, Polaroid camera, spectacles and address book amongst other things had been stolen, as well as my brief case with all my papers, except, luckily, the ship's papers and our passports. This was fortunate because I usually keep my passport in my brief case. The theft of spectacles and binoculars was a considerable handicap to me, but we sailed just the same although there was a forecast of fog. I had to rely on Peter Rowland in the Havre channel to tell me whether the objects I could see were buoys or small boats, but having got clear we set a course for the Nab Tower and put her on the auto-pilot. After about an hour we got into a dense dog which, in itself, would not have mattered; the engine driven generators were not charging and I had used a lot of power by running the gyro compass through the night; so I decided to return to port, as it was obvious that the radar could not be run for long. After turning round to return to Havre a heavy vibration was felt aft, so on berthing in the Bassin du Citadelles I asked the harbour authorities to recommend a diver for me to see what had gone wrong underneath the transom. The diver, an excellent Frenchman of the best type, found and cleared from our propellers an enormous polythene container such as are used for wrapping and transporting bananas from West Africa. These things and their like are becoming daily more of a menace to propeller driven small craft; these and lobster pots; so much so that it is really not worth making a fishing harbour in the dark unless one has a good searchlight.

We told the diver about our theft and he said we should report it to the docks police, as some of the articles stolen, such as the cushions, had a very distinctive pattern, also the binoculars and ash trays would be easily identifiable. He brought a police inspector to see me and I was asked to make a statement for their head office describing the articles stolen. After this *Procés Verbal* we sailed the next day and I managed a bit better, having been lent a pair of binoculars by our friend the diver. I hardly expected to see any of our stolen goods again, so my surprise can well be imagined when I received a cable from him saying he had seen some patterned cushions and leather ash trays with shot underneath, like ours, on board a British ship for whom he was doing some diving. The chief engineer, an Englishman, told him he had bought them from a former member of the crew who had left earlier on in the week. I found this man's address in Hull from the owners, and they gave me a forwarding address.

After talking to our local sergeant of police in Northampton I persuaded him to ring the C.I.D. in Hull to ask whether the gentleman I named had a record of crime. The answer came back; 12 convictions, mostly larceny. The C.I.D. at Hull were quick enough to go out that evening to the address given and found my binoculars, cushions and ash trays in a car. I was asked to go to Hull to identify the items mentioned, as well as a few more, but I never got back the rest of the things, including the Polaroid camera, my brief case, and an address book, or my spectacles. It has never been clear to me why the thief should have taken the spectacles unless it had occurred to him I might wake up and chase him, when I would be at some disadvantage in seeing him and identifying him without my glasses:—maybe he was right there.

A man in Hull was interrogated by the police after we had identified our belongings, and he was placed in custody on remand for nearly three months before the case came up; then he was let off by the magistrates. As I expected, he said he had bought these items of ours from a Frenchman who had offered them cheap in a bar at Le Havre.

The weakness of our case was that none of us had actually seen the thief at any stage, although there was an indication that I had frightened him off by switching on my light during the night to look at my watch. The lost ash trays, of special design with shot enclosed in a leather container, making them stable in a rolling ship, were a present from Gianni Agnelli, one of the most charming customers I have ever worked for. He bought and loved the first *Tramontana*, which he kept in Italy and the south of France. We also designed for him a 90 foot fast cruiser which was well built by Pietro Baglietto. She was called *Trenta*, and so far as I know he can maintain a cruising speed of 25 knots in almost any weather. Gianni is now boss of Fiat but still has time for a bit of yachting. He gave me six of these special ash trays, and it was one of these which had obviously been dropped by the thief on the cabin top in making his hurried exit. I was sorry for the police, who had acted so promptly and taken so much trouble over the case, and also had recovered much of our property, which could never have happened if they had not acted immediately our sergeant rang them.

The *Dimarcha* still gives us a lot of pleasure and as I write this we are about to set out from Hamble to explore some of the lovely harbours on the south-west coast of Ireland.

In 1971 we called in at the Isles of Scilly on the way to Ireland and were shown over the extraordinary tropical gardens of Tresco by the owner, Commander Tom Dorrien-Smith.

It is a dangerous thing to say, but now there are so many aids to navigation even in the Land's End and Scillies area, known as a graveyard for shipping, one should be able to keep clear of trouble so long as the engines keep running. We are equipped with Decca Radar 202 and a good D.F. set on which one can get the bearing of Round Island on departure from Glandore, and Mizzen Head on or before Land's End is abeam running the other way. If you want even more help you can have Decca Navigator, which in the Great Britain and Irish sea area can give you your position instantaneously. We have not fitted this assistance so far, though I think we shall soon, possibly in miniature form. At Mizzen head, besides a 200 mile range D.F. beacon, they have a "Racon" which shows up on a ship's radar p.p.i. as an arrow to distinguish it from the mass of shipping apt to be shown up on the p.p.i. at any given time. This arrow, which operates from a transponder located at the Mizzen Head station, also flashes the Mizzen call sign

MZ at regular intervals of 10 seconds to make identification positive.

In making a very interesting visit to this Mizzen Head station I was shown all the apparatus by a kind keeper, Mr. O'Regan, who told me an interesting fact about the reception of his T.V. set. Normally he can only get the Irish programmes but, with a steady high barometer (anti-cyclonic conditions) he can receive the B.B.C. transmissions from England quite well. This interested me because in 1969 when we were crossing in the *Dimarcha* from Land's End to Kinsale we received the whole B.B.C. programme covering the *Daily Express* International Offshore Power Boat Race when 55 miles out from Land's End in the Atlantic on our little Sony set placed on the floor of the saloon and with a 3 foot whip aerial. Perhaps both of us were getting it from a satellite?

There are a few minor alterations we could make if another *Dimarcha* was to be designed but nothing of a major nature.

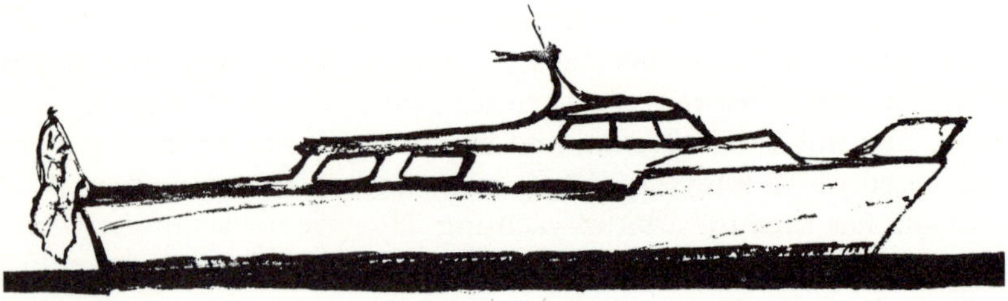

15

A Sales Effort which was near disaster

Only the other day I was sitting on the front at Cowes when to the horror of one of my companions he saw one of his brand new speedboats, recently acquired for the Prince of Wales, being towed by one of the motor boats belonging to the Royal Yacht. This was a severe blow as the boat—a well known make—was alongside the *Britannia* at the forward gangway with the engine cover lifted and men working on her as we had passed about half an hour previously.

This unfortunate affair reminded me of a time some 15 years ago when Prince Philip had been thinking of acquiring a speed boat for his own use

Fig 63 Fast Motor Boat Swordfish (built 1952)

so the Admiral of the Yacht, Sir Conolly Abel Smith, had organised a demonstration of two of our boats.

One was the twin engined *Swordfish* (Fig. 63) and the other a 19 footer with a single Gray engine installed, similar to that supplied to Loel Guinness for his latest yacht.

I went across the Solent to Cowes driving the *Swordfish* myself, and left Jim Hazel to come over after me and be ready to change over with me when I had finished demonstrating the *Swordfish*. The Duke was quite impressed with her, as I knew he would be, and of course I let him drive it and bring it alongside, which he did superbly. Then came the turn of the 19 footer, and I took over from Jim Hazel with the Duke sitting alongside me. After a few minutes Hazel, who was in the sternsheets, leaned forward to tap me on the shoulder and tell me the boat was sinking! Looking aft I saw that indeed the aft cockpit had a great deal of water in it, so I told Prince Philip about this and suggested we get back alongside the Royal Yacht as soon as possible.

The Prince just laughed and luckily we were not very far away and I got alongside again as soon as possible. Having reached the gangway and jumped aboard I told Hazel to go and beach her in Osborne Bay if he could not find out where the leak was coming from. The Prince kindly took me into his study and gave me a pink gin and after a short time I got a message to say that it was only a water cock left open on the service to the shaft log, and this had been filling the boat ever since we left Portsmouth. No sale resulted because I think some rival firm presented one of their products.

Technical Appendices

Appendix I Propellers

We had, of course, over the years carried out a great deal of development. It was only by building a craft and watching it carefully through trials and service running that we could be certain of improving the breed each time a new type of craft was embarked upon.

It was not until the argument about the optimum shaft speed of the *Braves* took place while we were in the feasibility study stage that I made a tour round all of the research establishments in Great Britain and Europe to discuss this high speed propeller problem. No expert propeller designer that I visited seemed to have done much work on the fully cavitating propeller, which was an obvious necessity in our case if we were to use shaft revolutions in the neighbourhood of 1,700. The most helpful of all the establishments visited were KMW at Kristinehamn in Sweden, where Lennart Pehrsson proved very knowledgeable as he had experienced the same sort of problems in dealing with the Royal Swedish Navy. He had not, of course, gone so far as we intended, our target speed being higher than anything his firm or anyone else had achieved up till then, and the power transmitted per shaft was much greater.

Lennart came up with a shaft speed of 1,200 after some consideration, but even this figure did not seem right to me, though at that time I had little more than a feeling that higher revolutions should be aimed at in order to cut down the size of appendages such as shafting, propeller brackets, intermediate brackets, if any, and rudders.

I had a hunch that much of the reason for rather disappointing speed results in, for instance, 510, the *Bold* class and the *Darks*, where the propeller revolutions were relatively low, was not due to inefficiency of the propeller but to the amount of appendage drag involved in the relatively large inclined shafting, propeller brackets, etc.

In the final resort I came to the conclusion the best thing we could do was to instal a cavitation tunnel ourselves, in order to carry out research on the problem of the high speed propellers and appendages, about which comparatively little seemed to be known. It was always my objective to attempt to test one of these high speed propellers on an inclined shaft to establish the difference between the efficiency in axial and inclined flow. There is still all too little precise knowledge on this subject.

During my tour around Europe on the quest for more knowledge of the high speed propeller I came across the excellent firm Kempf & Remmers in Hamburg who were making a speciality of producing apparatus required for research in naval architecture. So I placed an order with this firm for a cavitation tunnel suitable for our type of research which, of course, included amongst other characteristics the ability to test a model propeller in flow the equivalent of 50–60 knots. To help me with the specification, and to watch the building of the tunnel I engaged the services of Paul Rader as senior hydrodynamicist. He had come over from Germany just after the war and was employed by Dr. Gawn, mainly on theoretical work in our field of high speed propulsion; but as he was of German origin there was a limit to how much work of a secret nature he could be employed on at the Admiralty Experiment Works. Dr. Gawn strongly recommended Paul Rader and we had no need to regret our choice.

Of course there arose many other research problems in which Paul Rader's experience and knowledge proved invaluable. I had, for instance, given a pretty comprehensive research problem to KMW to test some high

Fig 64 The Vosper Cavitation Tunnel

speed propellers in their tunnel in axial and inclined flow and very interesting results arose from this.

Also at about the same time, at my special request, the Ministry of Supply, which was responsible for the high speed launches we were designing and building for it, gave us a contract to carry out a series of experiments to obtain a better idea, if possible, of the strains and stresses set up in the hull structure of these fast little craft when running at high speed in a seaway. The results of all this work, in which I was much assisted by Paul Rader and others, was eventually published as a paper ("The Planing Performance, Pressures and Stresses in a high speed Launch," by Peter Du Cane) and read before the Royal Institute of Naval Architects, when a very valuable discussion resulted. This paper certainly increased our all round knowledge of the problem although much more precise and detailed knowledge is required to this day.

But the most important function of Paul Rader in his initial days was to carry out a methodical series of trials in the cavitation tunnel from which we were able to design the propeller required to give the *Braves* their outstanding performance.

Without going too closely into the technicalities of propeller design it is perhaps sufficient here to state that in the normal sub-cavitating propeller both sides of the blade are fully wetted and it is probably true to say that the upper surface of the blade does more work than the lower face. This is considering the blade as an airfoil where the upper surface produces more lift by suction effects than the lower by pressure effects.

There comes a time, however, when it becomes impossible for the water to follow the profile of the blade back, when cavitation occurs. This cavitation really takes place when the pressure is reduced so much in trying to follow the shape of the blade back that it is reduced to the vapour pressure of water. Then the blade is no longer completely covered with water and a cavity or cavitation occurs over the region where the pressure has been reduced to the vapour pressure of water; in other words it has reached its boiling point over this low pressure area.

It is most undesirable in a propeller designed to be fully wetted, i.e. cavitation free on both sides, if cavitation occurs over a certain area of the back—usually in a chordwise direction and a collapse of the cavitation takes place further back on the blade. We then not only lose efficiency but, where the collapse takes place on the blade back, severe pitting or erosion takes place which can, and does, if left, eat a hole right through the propeller.

Cavitation can occur also on the lower working face when thrust and efficiency will be reduced, but it is rare for extreme erosion to occur from this cause. When we reach the point where it is impossible to avoid cavitation—and in the really fast ships it is indeed impossible to avoid this condition—then it is found to be better to use a fully cavitating or super cavitating propeller. Here, it is accepted that cavitation must occur on the back of the blade, so we design the section of the blade in such a manner that from leading edge to well past the trailing edge a large blister of cavita-

PARTIALLY CAVITATING AND FULLY CAVITATING SECTIONS OF PROPELLERS.

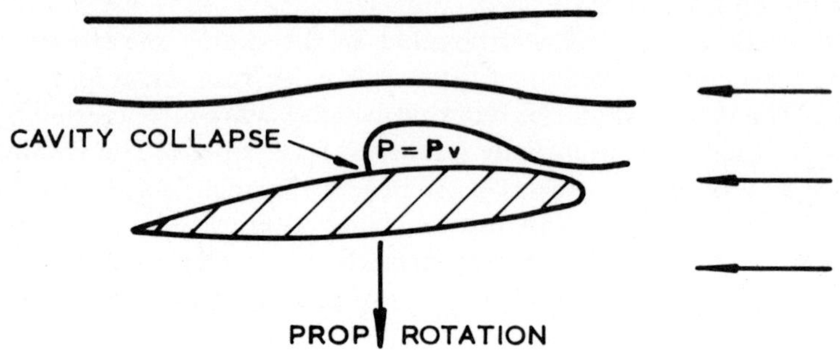

STANDARD AIRFOIL SECTION
PARTIALLY CAVITATING

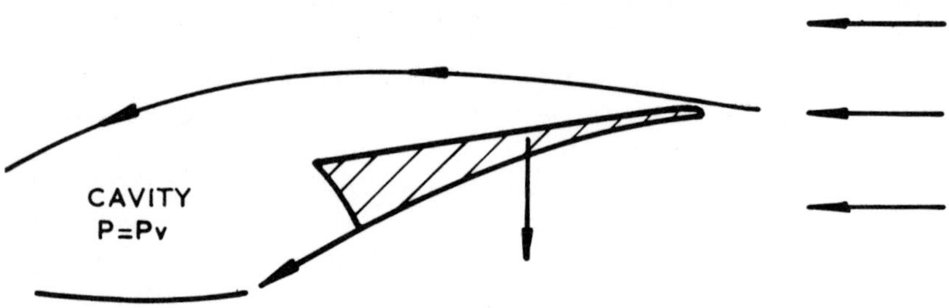

FULLY CAVITATING SECTION

P = LOCAL PRESS
P_1 = VAPOUR PRESS

tion occurs, which collapses downstream of the propeller altogether. This is all right so far as it goes as the lower portion of the blade will now do all the work and must be designed to do this in the most efficient and cavitation free manner. The state of affairs shown in the sketch opposite will help to explain the phenomenon.

Here it can be seen that there is no drawback, in fact some advantages, in allowing the trailing edge to have some finite thickness usually called a blunt trailing edge. This is very workable and a satisfactory state of affairs so long as the blunt trailing edge is within the cavitation blister. It should, perhaps, be mentioned at this stage that there is usually a penalty to pay for this state of affairs as the cavitation blister is quite likely to collapse on any rudder placed behind the propeller and within its slipstream. This can happen to a very severe extent in, for instance, racing boats or really fast patrol boats. The only really satisfactory remedy is to locate the rudder or rudders between the propellers and well out of the slipstream. A certain loss of manoeuvrability per unit area of rudder has to be accepted, although there are methods of improving this state of affairs not connected entirely with the shape or area of the rudder.

It has to be acknowledged that the fully cavitating propeller is less efficient as a propeller but the overall efficiency, when considered as a propulsor, is greater because less total drag is involved in the much smaller appendages required.

Despite only one side of the propeller blade doing all the work it is still of great importance to get the optimum shape and section, also the speed of revolution just right. Hence the reason for it being well worthwhile installing a cavitation tunnel at the stage we found ourselves in about 1954 when the discussion about the propeller speed of the *Braves* and other similar craft was being discussed.

On the subject of research and development much could be written but apart from giving an appendix listing the many papers and articles in the technical press, and reports published in my era, I will mention only some of the more important items.

Appendix II Broaching

It had long been a particularly interesting matter to me in the course of running the various craft which came my way that they were apt to behave in a rather uncertain and sometimes very undesirable way when running in a following or quartering sea. In conjunction with Professor G. J. Goodrich we prepared a paper called "The Following Sea, Broaching and Surging" which contained a good deal of useful information following some tests of a model hull of the typical semi fast type (i.e. not planing), actually it was our first corvette type built for Ghana, in the large tank at the Ship Division of the National Physical Laboratory at Feltham. The model was self-propelled though some restriction was imposed on roll angle and sway. If the more theoretical part of the paper is not of interest or quite comprehensible to some of my readers, at least I would strongly recommend a perusal of the various first hand accounts given by the authors and reproduced herewith:—

Examples of Broaching in Actual Ships

Early examples of this behaviour can be found in *Hakluytus Posthumus or Purchas His Pilgrimes*, published in 1625[1] in a chapter entitled, "A true repertory of the wracke, and redemption of Sir Thomas Gates Knight; upon and from the Islands of Bermuda." In this description is found the following passage:—

"Once, so huge a sea brake upon the poope and quarter upon us, as it covered our shippe from stearne to stemme, like a garment or a vast cloude, it filled her brimme full for a while within.

"This source or confluence of water was so violent as it rusht and carried the Helm-man from the Helme and wrested the Whip-staffe[2] out of his hand, which so flew from side to side that when he would have ceased the same again, it so tossed him from star-boord to larboord as it was God's Mercy it had not split him."

Then in 1764 we have: *A treatise on naval architecture founded upon Philosophical and National Principles.* William Hutchinson, Master.

"On scudding before a wind in a storm. Many precautions are necessary on this occasion . . . for it should be considered that a ship at such times may be so overpressed and plunged so deep in the sea, forward, that instead of dividing the water on each side in an easy manner the full parts of the bow, and other stops not designed to be in the water, drive a great body of water above its natural level before her bows which may increase the resistance ahead to such a degree, as will decrease rather than increase the ship's head way, and that in proportion to this rise of water; and the stern being lifted up as much as the head is pressed down, the helm must then naturally lose so much of its power at the same time; so that when the wind may be

[1] By kind permission of the National Maritime Museum, Greenwich.
[2] Whip-staffe was a form of vertical tiller performing the same function.

going four times as fast as the ship, and the waves twice as fast, it is not to be wondered at, that ships are sometimes broached to, against the power of the helm.

"The remarkable loss of the Sussex East India ship homeward bound from China in the year 1737, first arose from her being broached to, when sailing before the wind . . ."

Clipper Ships

In the days of the famous tea clippers as exemplified by the *Cutty Sark*, for instance, the behaviour of the ship in a following sea was of the utmost importance to the overriding necessity to make a passage from port to port in minimum time, in other words, in maintaining maximum speed. In the case of the tea clippers from China and the grain carriers from Australia, a voyage half-way round the world was involved.

Being square-rigged ships and unable to sail within say five points from the eye of the wind most of the voyage would be carried out from the Cape of Good Hope to Australia or China, and from the latter country to Europe by rounding Cape Horn in the violent Westerlies encountered in the South latitudes, usually known as the "Roaring Forties" or "High Latitudes".

To make best use of these always strong following winds the ship had to be driven as fast as possible, which, in a huge following sea, involved the utmost in skill and daring to keep the ship running as fast as possible and yet to avoid the dreaded "broach" which would be likely to take the masts out of her, as well as sweeping her deckhouses away, or at least severely damaging them.

At this stage it is perhaps appropriate to describe the typical behaviour of a fast ship in a following sea. To start with, it should be stated right away that with the sea right aft running exactly in line with the ship's longitudinal axis the motion is unlikely to consist in more than pitching due to the encounter period. However, the true following sea is a rarity owing to considerations of the course required to be made good, which may well not coincide with the direction of the wave system. Also the realistic sea encountered in the oceans of the world is seldom sufficiently unidirectional and long crested to avoid the occasional obliquity of approach.

So mostly we have, in practice, to consider a noticeable element of obliquity in the angle of encounter of the wave system which will give rise to disturbing forces leading to more or less violent yawing, which, in marginal conditions, if not checked by highly skilled steering, will be liable to develop into a wild and uncontrolled yaw through about 90° until the axis of the ship lies along the trough of the wave system. In this precarious position the wave to windward will be liable to break over and crash on board the ship with considerable force. The resulting damage, depending upon the type of ship, can amount to carrying away the deck structures, leaving the ship vulnerable to filling or, as in the case of the square-rigged ship, a loss of the masts due to the impossibility of reducing sail carried aloft in the available short time—a matter of a number of seconds only!

Even if the masts did not fail under the strain the ship would be knocked down with yards and canvas in the water to leeward.

It was here that the skill and vigilance of the man at the wheel counted much towards retaining control directionally in these towering seas, the waves of which would come up from astern and remain posed at the poop, giving every appearance of being about to break aboard. Most of these surprisingly enough did not, in fact, break, but so unnerving were these mountainous seas approaching from aft, that it was a very usual practice to rig a screen between the man or men at the wheel and the transom so that he could not see the nature of the threat.

In this case of the "Cape Horners" the avoidance of loss of control would consist mainly in good helmsmanship and watching the trim and sails. Too much speed leads to difficulty and even danger in directional control, the reason for which will be explained later.

However, it will be readily understood that after a certain point in the development of the storm and the resultant sea it would become too dangerous to consider coming round, even voluntarily, through the trough of the sea on the way to the course required for "lying to" on the wind. In fact, the ship in these circumstances would be likely to undergo the same process as for an involuntary "broach".

For this reason most of this type of vessel would "run it out", sometimes even to the extent of reducing to bare poles and streaming a long heavy mooring line or sea anchor aft.

Broaching of a Four-Masted Barque in High Latitudes

The story of the *Parma* provides an example of the behaviour in heavy following seas of one of the famous grain ships while running from Port Brongton near Adelaide to Falmouth rounding Cape Horn on the way.

A few short passages are quoted here which will suffice to describe the nature of the problem. For those who are interested the book can be thoroughly recommended: the phraseology used is reminiscent of that of Joseph Conrad.

"The wind cannot get much stronger. In a few hours, at the most, it should begin to ease a little. We shall have to go on now, whether we wish it or not. The Old Man has never heaved-to in one of these big ships. We have too much sail set to bring her into wind, even if we wished to.

"Then came the squall which almost robbed us of our morning, of all future mornings.

"With awful suddenness there came a mighty squall which made its predecessors seem like Doldrum airs. The helm would not go up, not for the combined striving of four boys and the mate: green water smashed along the length of her. . . .

"There came a giant greybeard out of the night at her, reared high into the sky; up, up she rose to it—not fast enough!

"Pressed down by the weight of water, blown on by her press of sail, she could not rise sufficiently.

"Down, down she sank poop under!

"She pooped. Of all the sounds that a storm-lashed sea can make surely there is none more terrifying, more expressive of evil exultation, than the lapping of solid green water over a running ship's poop.

"The water came fair aboard her over the stern, sweeping down the saloon and messroom skylights and down the charthouse companion, filling the saloon alley-way, washing out the mates' and stewards' cabins, sweeping overboard the compass by which the helmsmen steered.

"The ship was now in a serious predicament. It was impossible to con the ship from the standard compass. She did not poop again. She avoided that. But a giant sea, before we were able to do anything about the compasses, came roaring at her while others of its fellows held her down; it hit her a mighty blow, carting her off before anything could be done to counteract it. She staggered badly as if she were a living thing that had been shot. . . . She fell on her side in the trough of the sea. . . . Broached to!

"What could we do? We still had too much sail on her to heave-to, and first we had to get her out of the trough . . . out of the trough—if we could get out—we could not 'run' without a compass. . . .

"She rolled as if she would go over, lose her masts, smash in the hatches. . . .

"There came a chance, about 10 o'clock—over three hours after she had first broached to. Down helm! . . . She was a long length to bring into the wind, a big ship to lie hove to in the belly of the sea.

"She had broached to with her long bowsprit headed southwards towards the ice not so far from here; she fell into the great troughs with her yards squared and the wind on her starboard side.

"The captain straightened her, got her off, and hove to on the other tack—superb seamanship."

The interest to the naval architect in this realistic account lies in the confirmation of the general pattern of the behaviour of a ship, whether she be sail or power driven when labouring in a heavy following sea.

Here undoubtedly the speed of the waves will be considerably higher than the ship even though she is under a full press of canvas. Such oceanographic information as is available would lead to the conclusion that on the assumption this gale reached 10–11 Beaufort scale the waves would be 1,000 ft. long, which will indicate a crest speed in the neighbourhood of 42 knots.

Now there are obviously two main dangers in this particular situation. The one is "pooping", i.e. shipping green water over aft and flooding the ship from a vulnerable quarter.

The other is broaching-to. The ship is in fact more vulnerable to each class of behaviour the faster she is running, though certainly in the case of pooping it is by no means easy to explain exactly why this should be the case. It could, likely enough, be due to the effect of the ship's own wake due to her speed which upsets the equilibrium of a big wave in the incipient breaking conditions.

As far as the broaching is concerned the probable reason for greater liability to this danger associated with high ship speed can be explained by the longer time during which the ship lies in a potentially dangerous attitude on the forward slope of a great wave leading to substantial increase of speed due to surge forward and subsequent rather sudden retardation with perhaps loss of rudder effect.

Most experienced seamen would give it as their opinion that other things being equal this type of ship in a tempestuous following sea is best left to "run it out" under bare poles if necessary.

It must always be a dangerous performance to try and bring any ship, let alone a square rigger, round through the trough of the sea to heave to.

Conditions were different in the case of the *Parma* as she was caught with too much sail on her and one gains the impression that the ship's company could not manage to reduce sail sufficiently quickly in the appalling conditions prevailing.

Also, having lost her compass while pooping it would have been impossible to run it out as very accurate steering would be called for to avoid excessive yawing and broaching. In the case of the *Parma* therefore, she had little alternative than to attempt, and by superb seamanship on the part of her captain, to succeed, in heaving to.

It is rather interesting to note what the author says about the squall which nearly overwhelmed the ship, that despite the combined strivings of four boys and the mate they could not get the helm up. The reason for this is somewhat obscure because the ship, after being overwhelmed in this way, can hardly be supposed to have excessive speed on her—the only factor presumably which could make the helm heavy to move. In general one would imagine the effort on the wheel to be proportional to the square of the speed through the water and the effective angle of attack of the rudder relative to the flow. To some extent a strong effect might be predicted from leeway or drift at the stern as the rudder is unlikely to be balanced.

This is a special case where, for reasons given above, it became necessary to heave to.

Hertzogin Cecile

However, on the authority of Viscount Head, now High Commissioner in Nigeria, who sailed in her as a young man, the *Hertzogin Cecile*, a somewhat larger and faster four-masted barque, never at any time contemplated heaving to in the high latitudes but would run it out under much reduced canvas.

Other types of sailing vessel suffer from being driven hard in a following sea to the extent that their directional stability is much reduced and excessive speed can lead to violent behaviour in yaw and associated roll. A century earlier than the tea clippers the following sea did not concern the mariner to the same extent, as their ships were much slower and invariably hove to at night, which was probably the most dangerous time for a clipper being hard driven in the Westerlies.

Tzu Hang

However, in recent times there was an extraordinary accident in this part of the world to a small sailing yacht aiming to round Cape Horn, described by the owner and skipper in his fascinating book. Here the yacht was running before a tremendous gale with practically no sail and a mooring line streamed over the counter. The owner's wife, who was at the helm, noticed a sheer cliff face of water approaching and could do nothing to avoid being overwhelmed. The next she knew she was in the water and the yacht was lying in a water-logged state with the masts and coach roof carried away. Reconstruction of events and from evidence found in the overhead beams, it became clear the ship had turned a complete somersault.

This is an extreme case and bears no relation to directional stability or broaching, in fact it must have occurred with the waves from right aft. The type of wave which can be encountered in this part of the world, will go far towards explaining this behaviour. The accident to *Tzu Hang* occurred nearly 1,000 miles from the West Coast of Chile. They saved themselves by a miracle of fortitude and skill to sail into Aranco Bay near Concepcion.

Fishing Vessels

In the case of a fleet of small fishing boats working out of Concarneau on the Atlantic north-west coast of France, no less than 8 vessels and 70 men were lost in one gale during November 1954. The subsequent inquiry reported in great detail having questioned all survivors who could give information relating to the actual sequence of events leading to disaster.

In some cases, although surviving, the crew members had evidently experienced the set of circumstances leading up to a close shave from disaster due to becoming overwhelmed by the sea. The general conclusion was to the effect that in most cases, where an accident had occurred, the cause was "la fuite"—in other words, running for it to a home port to leeward, or at least trying to do so! There is a paradox here in that when broaching uncontrollably into the "trough" the ship tends to be overwhelmed, whereas one of the safest methods of heaving to suggested in this excellent report is to "tenir le cap en travers".

These small rather broad transomed craft in trying to run it out and make port to leeward as fast as possible had, in many cases, broached, and as a result became overwhelmed in the "trough". If the deckhouse is washed away the ship is open to the sea and will fill up and founder in very short order.

Broaching of a Battleship

Considering now the larger power-driven vessels, one author well remembers his first experience of broaching in an old battleship, HMS *St. Vincent*, during the First World War in a very heavy winter gale off the Norwegian Coast, in November 1917.

For some hours we had been running before an increasing gale force wind and sea, so that when it became necessary at the extremity of our patrol line to turn through 180 deg. on to the reciprocal course the quartermaster probably applied too much helm under the prevailing conditions. The result was a violent and uncontrolled yaw or "broach" into the trough of the sea, while at the same time the ship maintained an outward heel during the rapid turn due to the centrifugal effect of the high *CG* which is above the waterline, assisted by the location of the ship on the descending wave slope, which, as the ship approaches 90 deg. becomes a slope tending to produce an outward heel.

Suddenly there was a crash overhead as the wave broke aboard and crossed the ship, causing much flooding below.

Most of the ship's company were asleep in their hammocks so that on being awakened by the crash and finding the ship heeled and staying over, they thought they had been torpedoed.

Conte di Savoia

A report by de Santis and Russo describes the violent yawing behaviour of the Italian liner *Conte di Savoia* when running in the severe quartering sea generated in mid-Atlantic, where the Gulf Stream meets. This paper also describes the measures taken to improve this behaviour. It consisted mainly in increasing aft trim and adding deadwood or keel surface, as well as speed reduction when necessary.

Trawler Behaviour in Following Sea

A somewhat similar state of affairs is described by W. Möckel in a paper on the seaworthiness of trawlers in *Fishing Boats of the World*, Vol. I. Here he describes a personal experience aboard a trawler off the Orkneys in gale force 9 from astern ". . . suddenly the ship was struck with great violence by a breaking wave coming up from astern which flooded the low foredeck nearly to the level of the rail. . . . Her behaviour conveyed the feeling that stability and buoyancy were exhausted and that she might at any moment capsize and sink."

Here again it will be seen that the correct procedure is to slow the ship so that the relative speed between ship and the overtaking waves is as much as possible while maintaining a "stern on" course relative to the advancing waves. This again requires sea room, but Captain Möckel makes the point that he did not consider it possible to attempt a turn through the trough of the sea towards the eye of the wind, though this was partly owing to the already reduced stability arising from a breaking sea having flooded the decks up to the gunwales.

An Extraordinary Roll

There is the case quoted in the *Naval Review*, extracted from the journal of the U.S. Naval Institute, of a Portuguese destroyer which had gone to the rescue of the survivors from an American merchant ship sunk by enemy action in the vicinity of the Azores in 1943.

After effecting the rescue the captain found himself about 135 miles from a port in which he could find shelter. He was informed of a forecast of increasingly strong winds up to and beyond gale force, so that he decided to run for port down wind in the hope of entering harbour before darkness. The course was 73 deg. to Ponta Delgarda and the wind backing to south-westerly, so the wind was on the starboard quarter.

At first he maintained 18 knots in an increasing following sea. From then on he increased to 20 knots, though steering was very difficult. Then he increased to 22 knots when the ship started to labour to a substantial extent. In the words of the captain steering became more tricky every minute. He therefore decided to try to "get away from the seas" by increasing speed to 24 knots. For a short time matters were improved and the ship seemed to be gaining on the waves.

However, as the fury of the gale increased the speed and length of the waves increased and the situation grew worse. The ship began to yaw heavily more than 40 deg. and gave the impression of a surf board. They now ordered speed for 26 knots, again with the idea of gaining on the waves, but this did little to improve matters.

Eventually as matters became even worse with the increasing fury of the gale the captain decided to reduce speed drastically, but he was not in time:

"Suddenly the ship was taken by a great wave and began such a yaw to starboard that not the whole force of the helm could stop her. She heeled slowly over to port more and more. . . . I gave the order stop port engine and finally stop both engines as matters did not seem to improve and we were practically lying down on the bridge with our feet braced against the bulkheads.

"The ship continued slowly but surely to heel further to port. We remained thus for a few awful moments until, at first slowly, and then more quickly, the ship righted herself again."

It was after this that the engine-room gauge had registered the terrific roll of over 67 deg.

This was unquestionably a case of broaching in a following sea, which was all the more violent for having occurred at a ship speed somewhat in excess of wave speed. The description of the ship "surf boarding" is interesting in view of the behaviour described in this paper and observed in the Feltham tank at N.P.L. in the case of a similar type of craft and in terms of speed increase is of rather sensational magnitude. This condition is very likely to occur as a preliminary to broaching. It will be realized that little is known quantitatively about this phenomenon in full-scale ships, as it is almost impossible to measure in the open sea or ocean.

Of course the roll recorder here is not, of necessity, able to make an accurate record of the roll owing to the effect of various extraneous athwartships accelerations likely to arise in such a case where very violent yawing takes place as well as rolling, and presumably pitching. To mention one quite important effect is the centrifugal force $mr\omega^2$ involved in a violent yaw.

This ship would possibly have been quite comfortable—at least comparatively so—if speed had been reduced earlier, thus allowing waves to overtake the ship at relatively high velocity.

Most authorities claim that the state of affairs when the ship is proceeding in a following, or nearly following (slightly oblique) sea at the same speed as the wave crests, represents the worst condition for directional instability.

There is, however, some evidence to the effect that if in fact this speed can be exceeded by the ship, then the liability to violent yaw and broaching will be greater, certainly until such time as the ship is going so fast that the following sea becomes in effect a head sea, as it appears to an observer aboard ship.

This is a hypothetical state of affairs as applied to normal types of ship.

Two First-hand Accounts of "Broaching" from a Yachtsman who is also a Naval Architect

It may be of interest to read the verbatim account of broaching given by Howard Chapelle, naval architect, who is curator of The Division of Transportation at the Smithsonian Institution in Washington D.C.:

"Broaching, or a tendency to broach is probably more common in power boats than we realise. I have had only two experiences in person, both mild but startling. In 1926 another fellow and I were running into Caesar's Creek, an inlet between the sea and lower Biscayne Bay, Florida. We had an old 26 ft. keel sloop, a former sponger, drawing 4 ft. 3 in. aft, and 2 ft. 6 in. forward, jib-headed mainsail with a rather long boom. The wind had been a fresh gale from the east-south-east for two days and was moderating. There was a sea about $4\frac{1}{2}$–5 ft. high I should guess. We started in, main with one reef in it. For the first mile we had no trouble. I observed as we approached the narrow part of the inlet that the seas were noticeably closer, crest to crest, than before. Suddenly we seemed to over-run a sea, the bow dropped into the trough and then buried. She ran into the sea ahead and suddenly I could not steer her; she turned sharply to port and the following sea over-ran her on the port broadside. She then knocked down but did not capsize. We had about two feet of water in the hold, however. Before the second sea hit her she fell off a little and I was able to get her back before the wind. The whole thing happened very fast and as my mate had cut the mainsail halyard we retained control long enough to get through the inlet.

"The second case was less violent—I was in a 30 ft. power cruiser at Ipswich, Massachusetts in the fall of 1938. We were running before a rolling sea into Ipswich Bay, to enter the river there. Just as we passed over a roller it broke rather heavily, and the stern seemed to rise very suddenly. The bow must have buried but I could not see as she was raised deck forward. She took a marked heel and a sheer, and ran off it about 40 deg. to her course, against her rudder.

"I do not recall whether or not we had overrun a sea but we suddenly went out of control to a marked degree. I reversed the engine and she came under control again. I did not consider the sea dangerous. The boat had almost no drag to the keel, a long and very sharp entrance and a broad transom. When the stern lifted the boat felt as though she had lost stability. This boat had been built from a half-model. I inspected this model later and it was much like a Nova Scotia Cape Island launch, the bearings being all aft. The Cape Islanders have been reported to have been lost by broaching in severe conditions in this manner and for this reason I believe drag to the keel is a 'must' in fishing launches.

"I have heard a number of descriptions of minor broachings but so often the observations were poor and one could not follow the sequence of events. I think that two things may happen—overrun a sea and hit the one ahead, bury and broach. The other is for the stern to be thrown up on a crest so the rudder is no longer in solid water. That is why beaching boats do not depend upon rudder alone—they have a steering oar at hand."

Behaviour of Surf Board in Breakers

There is the example of the behaviour of a surf board in a steep wave about to break which can be termed a limiting case of following sea behaviour which can lead to the equivalent of broaching.

In the case of a wave, preferably of the long-crested variety, resulting from a heavy swell approaching a shallow beach, especially if about to break when the particle velocity in the crest approaches the velocity of wave translation. At a certain point the wave starts to move forward in the form of an orbital circle at the speed of translation—at least the top part of the wave moves at this speed though in the trough a considerable velocity towards the wave is generated in a direction opposite to the velocity of the crest.

The art of surf board riding consists in placing the board so that the aft part runs in the crest where is maximum velocity and it is inclined slightly downwards so that the forward part operates in water at reduced velocity as it becomes further from the crest.

In this way a sort of equilibrium is maintained so that the aft part is attempting to push the board ahead faster than the forward sections will allow. In other words the board moves through the water because it is perpetually sliding down hill.

If this situation is allowed to continue too far inshore there may come a time when the forward sections can be actually operating in a reverse flow so that some considerable tendency for the board to broach into the trough will be observed.

Certainly it is not difficult to understand that this differential in the velocity of the water particles in the actual crest and those somewhat further ahead towards the trough cause an unstabilising moment leading to broaching of the board.

To reduce this tendency either the board must be small in length compared to the wavelength, or the wavelength must be relatively large. As to carry a human, there is a minimum size for the board, it can readily be understood the best conditions for directional stability are created by very large waves and by using boards as short as are practicable.

Almost flat beaches are desirable for this sport because only under these conditions do the waves remain in the state of incipient breaking for long.

Finally, of course, after the wave has broken in the shallow water adjacent to the beach a shallow flow will continue forward at the crest velocity and then return at the trough velocity.

Appendix III Roll Damping

One of my lifelong friends and a very good customer of ours, Loel Guinness, after having two ex-naval craft converted to yachts by us, suggested one day that there was no anti-rolling device such as had been fitted in larger ships available for the medium sized yacht. He inferred that he was really interested and that if I could demonstrate to him on my yacht the *Sea Victory* a practical unit capable of ironing out the worst of the rolling, which most of these motor yachts suffer from, he would, in turn be pleased to give me an order to fit an installation of our design in his yacht the *Calisto*. This seemed to me a worthwhile exercise because, apart from other considerations, we were badly needing another activity to make proper use of our engineering and other facilities which had been somewhat short of employment since the war.

Naturally, I made a close study of all that was available on the market at the time; this really consisted of the Denny-Brown retractable gear, which had been developed during the war largely at Admiralty expense, and the Sperry gear which was developed in the U.S.A.

Though a very worthwhile characteristic and possibly essential in a number of ships, I came to the conclusion that if we could avoid retraction and arrange for a fin operated from mechanism within the ship, but which was nevertheless fixed in its outboard position and not capable of being retracted, we should save a great deal of complication and space. This latter feature, in itself, would enable many yachts and small ships to be fitted with roll stabilisation apparatus which would have been impossible if the fully retracting type had been the only type available. In fact most ships or yachts already built would have found it impossible to fit anything at all without major internal alterations to machinery or other spaces.

Study of many of the hulls of the faster type of warship or yacht disclosed the fact that it was quite possible for a fin of low aspect ratio to be fitted in such a way that it could fulfil its function of damping the roll movement while remaining within the rectangle formed by maximum beam and maximum draught. This meant that the ship so fitted could go alongside a quay or even touch the ground without damage to the fins.

Needless to say, the actual anti-roll torque is supplied to the ship by twisting the fin from inboard so that an angle of incidence to the flow over the fin, which would provide the necessary force which, multiplied by the effective distance between fin centre of pressure and rolling axis of ship, would supply the required moment in opposition to the rolling moment set up on the ship by the influence in the first instance of waves.

This system represented a considerable simplification, and owing to the relatively small and easily installed mechanism within the ship seemed to be just what we wanted. It would also be relatively cheaper.

In order to obtain an idea of the optimum plan form and section within the limitations we felt it was necessary to impose, we asked Professor Terence Nonweiler, an aerodynamicist, to make an investigation and submit to us a report on the optimum configuration of fin and its section, with the object of obtaining maximum lift or force from a given area which so far as outreach was concerned was limited by the shape of the ship. In most cases an aspect ratio of one or less would be indicated.

Professor Nonweiler came up with a suggestion that we should use a double split flapped configuration, but this proved to be rather unreliable and a bit of a nuisance when fitted to our *Sea Victory*, so with the advice of the Professor we changed to an unflapped fin of symmetrical NACA section turning through 30° each side of zero, instead of the 20° envisaged for the split flap type. By increasing the angle to 30° very little, if any, lift was lost and the general simplification made it well worthwhile.

The system worked well but it should be acknowledged here that one of the most important features in setting up this installation consists in the control of the fin movement and here we were much helped by the advice of John Bell, then chief research engineer of Muirheads Ltd. who supplied our control systems for many years and thereby relieved us of at least one problem in a not particularly straightforward piece of engineering development for a shipbuilder.

While testing the split flap arrangement on the *Sea Victory* my family were quite often with me and a lot of rude remarks were made when we were stuck with one fin jammed full over in a seaway and had to make harbour with quite a list. However, with the exception of an occasional burst hydraulic hose and such like awkwardnesses, the development went very well once we had the simple unflapped fins, and it was not a long time before we felt sufficiently confident to offer to fit them to Loel Guinness's yacht *Calisto*. He made a deal with us that he should be repaid by way of 10 per cent royalty on subsequent installations until the cost of his installation was repaid. Incidentally we called them "Roll Damping Fins" because before any movement of the fins could take place there had to be a movement of the ship so the fins were, in fact, dampers to the motion initiated by the waves. A true stabilisation would envisage a signal being sent to the fins from an approaching sea disturbance so that they could be put into such a position that the moment set up by the fins would anticipate and counteract the effect of the approaching wave disturbance. This would involve a control system which has not yet been produced and in any case would be very complicated and expensive even if practical at all.

The control in our system is sensitive to roll velocity; in other words, there has to be a roll velocity on the ship before the fins can be moved to counteract it. In the more sophisticated systems the roll velocity signal from a rate gyro can be mixed with an actual roll amplitude signal which can be of value when the ship is heeled over but not moving rapidly in the rolling axis. This state of affairs is quite apt to occur in a following or quartering sea. It is then that for some considerable time a steady angle of heel will occur and at the same time an almost, or quite, non-existent rolling velocity will be felt by the ship. It should, perhaps, be explained that the simple gyro sensitive to roll velocity consists of a gyro whose axis lies athwartships and whose deflection or precession will be proportionate to the roll velocity.

I am glad to say the roll damping fins in the *Calisto* were a great success and from this installation we went from strength to strength fitting, amongst other yachts, the magnificent *Christina* belonging to Arie Onassis. This was indeed a challenge compared to anything we had attempted up to date and, very sensibly, the owner allowed us four days in this magnificent yacht all to ourselves. Our party consisted of Peter Shepherd, Jack Pacey and myself, and at first things looked a bit black, as one of the actuating rods was found to be bent and we had to work for many hours in the heat of a workshop just over the boilers to straighten it up. However, by the time this work was completed and the installation working again, a nice convenient mistral blew up which enabled us to demonstrate to the captain and crew the very effective installation with which they had been supplied.

Incidentally, this was our first four fin installation as a two fin installation would have involved fins too large for the ship; they would have extended beyond the maximum beam of the ship and so been a menace when going alongside a quay or another vessel. In this case we elected to fit each of the fin installations with its own independent gyro control so that no rods, levers, or linkage were involved running through and across the ship.

Fig 65 S.Y. *Christina*

One of the best ways of telling quickly if the roll damping system is likely to work satisfactorily in a seaway is to force roll the ship. This consists in reversing the leads from the control gyro so that instead of the fins being twisted into the incidence for damping the roll the opposite takes place and there occurs a gradual build up of rolling until a maximum is reached; this is sometimes as much as 30° each way, which is a big roll at any time and unpopular with chefs whose pots and pans go flying all over the place. Anything much less than 25° each way is not much use. It so happens that we asked permission to do this in the *Christina* and most of the ship's company were warned, excepting the maid who was cleaning the owner's Roman style bath; she slipped and fell in head first but emerged wet through but smiling.

Another interesting job which arose at this time was the fitting of an installation, also consisting of four fins, to the yacht of a famous tanker owner, Mr. Ludwig of Bulk Carriers fame. His yacht, MY *Dangin* was built in Japan at Kure in a shipyard leased from Japan by Mr. Ludwig. This shipyard has gone back to Japanese ownership since this time. At Mr. Ludwig's express wish I travelled out to Japan in order to witness the original trials and take a ride in the ship from Kure to Hong Kong where the joinery, cabin fittings and decoration were to be completed. Just as we were ready to start out from Kure a typhoon blew up which kept us in harbour for a few days. When we did eventually steam out into the open sea most of the worst of the typhoon had disappeared leaving only a swell, which though not very violent was enough to test the roll damping equipment. Peter Shepherd, who was representing us in charge of the installation, and I helped out a rather small ship's company by keeping a night watch each during the four days and nights of the trip.

An interesting feature of the *Dangin* was that she was fitted with anti-pitching fins at the bow which had been suggested to the owner by the Davidson Laboratory. These fins, which were attached to the forefoot somewhat like the configuration of the wings of an aircraft, were intended to reduce pitching and slamming in a head sea. Unfortunately we soon discovered that the effect of these fins was to make the ship very unstable in the yawing axis. This is not difficult to understand because with even a slight amount of heel such as would be caused by the use of the rudder the tendency will be for the anti-pitching fins to present a noticeable fin effect acting as would a fore and aft fin placed forward. The result was extreme instability in yaw and therefore great difficulty in steering. A small amount of helm applied would lead to an uncontrollable turn which then had to be counteracted with large movements of the helm in the opposite sense, and before any correction could be completed the ship would be swinging away too much in the opposite direction.

In other words she was behaving as a directionally unstable ship. Strangely enough, because they are not often good in this sense, the autopilot was able to hold her steady. This was of the American Sperry type which is controlled with a rate term as well as a term directly proportional to the difference between ships head and course desired. This rate term appears to make all the difference to the effectiveness of course keeping and it would really have been an impossibility to have had to steer manually for any length of time. It happened that the skipper asked me to steer the ship in through the rather difficult entrance to Hong Kong, and I had my work cut out to avoid her sheering all over the entrance. Removal of the anti-pitching fins in Hong Kong put all this trouble straight and the ship became directionally stable and easy to steer.

In the course of my close association with all the problems relating to roll damping I came across many interesting, unusual and sometimes amusing problems.

One of the most interesting related to the order we had received to fit some of the *Leander* class frigates with our roll damping fin installations for the Royal Navy.

All went well, or reasonably well, as it has to be admitted there were one or two matters which required rectification in these prototype installations, which were probably the largest and most powerful we had yet produced. Unfortunately we had to change the spherical ball joint at the head of each cylinder rod in the hydraulic operating system and the hydraulic control valve proved to be too sensitive and apt to cause a somewhat jerky movement. This was really because the valve controlling the flow of oil to the cylinders was too large in diameter so that only a few thousandths of movement would allow a sudden surge of oil to the cylinders, resulting in a jerky movement. The valve was what we call unstable and it would have been better both to have a little chamfer at the lips of the valves and to adjust the flow to the cylinders in a more controlled manner by making it necessary to move the valve through a greater distance for the same flow of oil. In other words, the relationship between valve diameter and travel should be arranged so that the valve diameter was smaller and therefore the valve had to travel further for a given flow. This was eventually arranged and the resultant valve operation became smooth while controlling the amount of liquid required to operate the cylinder smoothly.

These matters were not difficult to put right, although expensive and tiresome owing to the delay caused.

Another rather more serious criticism was levelled against the Vosper fins which related to operational matters. The prime function of the *Leander* frigates is an anti-submarine role, so it is essential, in order to obtain the best results from the Sonar, that the minimum disturbing noise emanates from the ship itself whether internal or external. One of the ships fitted with our fins reported that when operating they made considerably more noise than the other type of fins fitted in this class of ship. There did not seem to be any very good reason for this and frankly we were sceptical, and so were a good many others of those concerned. It seemed more than likely that the propellers would be making far more noise than the roll damping fins—certainly when anything approaching maximum power was being transmitted to the propellers.

Eventually, as this question did not seem to be soluble any other way, the Fleet Anti-Submarine Officer organized a trial which promised to settle the matter once and for all. Two *Leander* frigates, the *Sirius* and *Arethusa*, were allocated to run trials in the Channel with a submarine submerged to periscope depth and listening. One frigate was fitted with Vosper fins and the other with Denny-Brown. Vosper were asked if they would like to send a representative in the submarine and I elected to go as it seemed likely to be a very interesting and important exercise. The other two on board the submarine, HMS *Talent*, were the Fleet anti-submarine officer, Commander J. J. Streatfeild-James, and Steve Lankester, chief scientific officer from the Haslar Tank, whom I had worked with on many previous occasions.

We embarked on board the submarine about 2100 hours at Portland and soon cleared harbour bound for mid-channel to carry out various exercises before our special trial became due to start at 0400 hours the next morning. We were given bunks to sleep in until our trial was due to commence, but comfortable as was my bed it was not possible to get much rest because every time the ship dived or surfaced what seemed to be half the ship's company filed past my berth bumping in to me. The only time which seemed at all peaceful was during the time we had dived to 100 feet and were running on the comparatively silent electric propulsion motors.

However, when our trial was due to commence I was given a pair of headphones to listen to the noise made by the various ships in passing. To start with there was little sea so that the fins in both ships were doing comparatively little work. The only noise we could hear quite plainly was that of the propellers in each ship. Eventually, in order to make quite certain the fins would be full over I asked the Fleet Anti-Submarine officer if we could ask each ship to turn away under full helm when they were level with us. It should be explained that most of this class of ship and certainly the *Leanders* would heel to a considerable extent under full helm—certainly enough to cause the fins to go to their maximum incidence position in an attempt to counteract the heel. As I expected, though the fins must have been hard over (and this we checked by communication with the ships concerned), there was absolutely no noticeable difference between the noise made by either ship. In other words, the propellers were drowning all extraneous noises such as, for instance, roll damping fins or even rudders, which can make quite a high degree of noise when full over. This is partly cavitation, partly separation and probably a certain amount of aeration.

All were agreed that whatever noise the fins might be making was not noticeable, even when they went hard up or down as the ship turned.

It was rather a thrill to be taken ashore by a naval helicopter from Portland. While the helicopter was holding its position just overhead of the submarine's bridge a cable was lowered with a harness which we had to put under our arms, when we were hoisted up to the helicopter in turn.

The result of this trial was entirely satisfactory to us and no more was heard about this trouble.

Appendix IV Directional Instability Problems in Fast Planing Craft
The design problems presented by all fairly fast craft are liable to take time to solve, though probably they are on the whole simpler than similar problems arising in aircraft. However, when dealing with the dynamic problems presented by the behaviour of a fluid, whether it is air or water, one is apt to be presented with some quite unexpected problems in behaviour despite every precaution.

This can involve much thought, time and money on the part of those concerned; however, more of these problems are discussed as they arise. While undoubtedly the behaviour of the fluid flow over the planing surfaces can make or break a fast planing craft, it must be realized that such matters as the fore and aft trim can make just all the difference—as well as the design and location of the rudder(s). One of the most imponderable problems which at this time (August 1968) we have not yet been able to solve, excepting to some extent empirically, is the effect of a fully cavitating propeller on the pressure distribution under the bottom in way of the propellers. In short we have discovered that the fore and aft propeller location can drastically alter the behaviour of the boat, and if we locate the propellers rather too near the transom an undesirable positive pressure is built up in the aft sections causing trim by the head and the inevitable liability to broach without much warning at, or nearly at, full speed.* This is an alarming experience and in a small boat can lead to ending the wrong way up. I have only touched on the subject here because again there are so many variables, but to give an idea of this rather baffling problem, we met with a most unexpected problem in the Danish FPBs three years back when trying to eliminate some unusual vibration in the aft peak above the propellers, as this vibration was doing bad things to the steering gear located in this compartment.

One standard method of eliminating vibration is to increase the number of propeller blades to produce the thrust required. In this case, without any other alteration, we produced a set of five bladed propellers to do the job. Greatly to our surprise we found the trim reducing to the danger limit (about $1\frac{1}{2}°$ above static) at a speed little more than 45 knots against a perfectly well behaved full speed with the three bladed propellers of 54 knots. The well known behaviour occurred, which commences with an unresponsive behaviour at the steering wheel combined with a slight list and tendency to turn slowly but surely one way—usually into the wind which is an important

* Since this was written intensive study and trials point to the fact that aeration or the drawing of air from the surfaces is likely to be the main cause of this build up of pressure under the aft sections of the bottom and amongst other cures sealing the hinge of the transom flap is helpful as well as increasing the flap chord (length).

factor—but it still happens in a flat calm. If one should persist in this condition the typical and very unpleasant fast broach through 90° will result combined with the even more unpleasant leaning outwards in the turn. Personally, I have had so much of this behaviour in various fast boats that, so far, I can always feel through my hands on the wheel and the soles of my feet when it seems likely to happen and take appropriate measures which will usually consist in slowing down. There are other ways of dealing with the situation by putting the boat into a really steep banking turn using plenty of helm one way or the other; strangely, away from the list is the more effective. This technique, once one is pretty experienced at it, will work for such things as demonstrations, where there is enough sea room, but of course is no use for an operational boat, and is only a short term palliative. To give an idea of the unpleasantness, not to mention danger involved, we had two of our engine room staff literally thrown over the rails into the sea on a recent occasion when such a broach occurred. It should be added that the so called warped planing bottom is more prone to it than one with nearly constant deadrise up to half way forward along the hull. Mostly this arises because with increasing speed the warped form will have a more rapid tendency to reduce the trim and immerse the forward sections. This phenomenon is discussed in much more detail in my book "High Speed Small Craft" (3rd edition, Temple Press).* As above stated we are only beginning to understand this behaviour and more investigation is required. However, we have so far always cured it throughout the speed range in all high speed craft delivered to a customer for operational use—in all cases by *ad hoc* methods, which can be a lengthy and expensive process. We know we can now design to avoid this phenomenon given time to test thoroughly and usually we are successful, but it is surprising how quickly we can run into unexpected trouble with some little suspected alteration. Many, in fact all, monohull racing boats which can exceed 70 m.p.h. can suffer from this trouble, but it is usually called "spinning" in the racing world. Hence at least one good reason for the increasing popularity of the multi hull.

It has since been established that by drawing air from the surface a propellor will increase the pressure in its vicinity.

* Now out of print. Another edition is in course of preparation for publication by David & Charles of Newton Abbott.

RESEARCH NOTES & PUBLICATIONS BY AUTHOR—1938–1968

Date	Title	Remarks
1938	Sixty-Knot Torpedo Boats?	Appeared in "Blue Peter" under pseudonym of "Swordfish"
1938	Future of the Motor Torpedo Boat in Naval Warfare	Appeared in "The Pelican"
1941	Some Notes on the Design of Seagoing High Speed Craft	Published in book form for private circulation and in serial form in "Motor Boating". (U.S.A.) 1941/42
1951	High Speed Small Craft—1st edition	Published by Temple Press
1956	—2nd edition	
1964	—3rd edition	
1952	"Crusader"—The Design and Construction of John Cobb's Jet-Propelled Boat	Special article for "Motor Boat & Yachting"
1954–5	Boatbuilding Plant and Facilities of Vosper	"Ship & Boat Builder"
1955	Warship or Weapon Design and Developments	
1956	Rolling in Small Power Craft	Article
1955–6	Reduction of Rolling in Small Craft	Technical Brochure
1958	Ship Behaviour in a Seaway Parts I and II (A commentary on recent research methods)	Special article written for Shipbuilding and Shipping Record following the International Symposium held at Wageningen September 1957
1958	High Speed Propulsion Research	
1959	U.S.S. *Skipjack*—The Shape of Ships to Come	"Engineering"
1959	Rolling Trials in the Wake of a Typhoon	"Engineering"
1960	Third Seminar on Ship Behaviour at Sea Held at Davidson Laboratory—Stevens Institute of Technology 13–17 June, 1960	Article written for Shipbuilding and Shipping Record
1961	Research in a Small Shipyard	Written for International Design and Equipment Number of Shipbuilding and Shipping Record
1961	In the Bay of Biscay (Roll Damping for the "President Auguste Durand")	"Motor Boat & Yachting"
1962	A High Speed Craft built to win a Race	Written for "Engineering"
1962–3	High Speed Craft	for Chambers Encyclopaedia
1963	A Craft to Beat the Nuclear Submarine	Specially written for "Engineering"
1968	Consideration of Underwater Forms as Applied to High Speed Craft	Written for Italian Enciclopedico del mare
1968	Gas Turbine Propulsion in Small High Speed Craft	Written for Italian Magazine "Mondo Sommerso"

REPORTS ON RESEARCH AND/OR TESTS CARRIED OUT BY OR TO THE ORDER OF VOSPER LIMITED (INITIATED BY THE AUTHOR)

Date	Title	Carried out by
1952	Tests on two Models of Vosper Launches	Saunders-Roe Ltd.
1955	Resistance Experiments on Twin Screw Motor Yacht	National Physical Laboratory
1955	Report on Cavitation Tunnel Tests of Propellers with inclined shaft in conjunction with Appendages	Vosper Limited Authors: P. Du Cane and H. P. Rader
1957	Further Tests on Vosper models 45 and 49	Saunders-Roe Ltd.
1957	Tests on Wooden Box Beams	Saunders-Roe Ltd.
1957–8	Special Study of The Whirling of Shafts	Vosper Ltd. (Author: J. R. Braham)
1957	Report on a Series of Tests Carried out on an R.A.F. 68-ft Rescue/Target Towing Launch to investigate Operating Torque recorded with Rudders of varying percentage balance	Vosper Ltd.
1958	An Investigation of the Calm Water, Head and Following Sea Characteristics of Four High Speed Vosper Hull Forms	Saunders-Roe
1958	Seakeeping tests with two models of high speed motor-boats—models 1766Z/1767Z	Ned. Scheepsbouwkundig Proefstation—Wageninge
1959	Model Tests of two planing forms and a Rounded Bottom Form in an Irregular Head Sea	Davidson Laboratory Stevens Institute of Technology, U.S.A.
1959	Resistance Experiments on Model Hulls 4031 A and B. (Trawlers)	National Physical Laboratory
1959	Resistance Experiment on Model Hull No. 4082 (Twin screw high speed launch)	National Physical Laboratory
1959	Report on Demonstration on Roll Damping Trials carried out in S.A.S. Oosterland and S.A.S. Rijger	Vosper Ltd.
1960	Hydraulic Pump Test Rig to Evaluate Noise Level in M.Y. *Dangin*	Vosper Ltd.
1960	Tests Over Waves on Vosper Model 45	Saunders-Roe Ltd.
1960	Cavitation Tests of K.M.W. Model Propeller in Axial Flow and in 12.5 Inclined Flow	Vosper Ltd.
1960	Tests in Calm Water to Investigate the Transverse Stability of Vosper Hulls 49, 55 and 70 (Interim)	Saunders-Roe Ltd.
1961	Report on Model Tests with a Fin Stabilising Equipment for SS *Olympic Eagle* (Tanker)	Ned. Scheepshóuwkundig Proefstation—Wageningen
1961	Report on N.A.B. Propellers fitted to M.Y. *Mercury*	Vosper Ltd.

Date	Title	Carried out by
1961	Lateral Stability Tests on $\frac{1}{35}$ Scale Model of "Ferocity" in Cavitation Tunnel	Vosper Ltd.
1961	Comparison of KA ME WA Fully Cavitating Propellers with Fixed Pitch fully cavitating Propellers	Vosper Ltd.
1961	Hydrodynamic Characteristics of 0·5 and 0·8 Aspect Ratio Roll Damping Fins with 0·3C Split Flaps	Vosper Ltd.
1961	Comparative Tests of Propellers for High Speed Craft in Axial and in 12·5° inclined flow	Vosper Ltd.
1961	Model Tests to Determine the Performance of Vosper Hull 70 in Calm Water and Model Tests to Investigate the Transverse Stability of Vosper Hull 71	Westland Aircraft Ltd. Saunders-Roe Division
1961	Report on Trials Carried out on 68-ft R/TT Launch HMAFV No. 2760	Vosper Ltd.
1961	Report on Strength Calculations for PC(H) Propeller	Vosper Ltd.
1962	Model Tests to Determine the Resistance and Optimum Attitude of Roll Damping Fins Fitted to Vosper Hull 72	Westland Aircraft Ltd. Saunders-Roe Division
1962	Fin Stabilisation tests for a fast torpedo boat Model 2277Z	Ned. Scheepsbouwkundig Proefstation—Wageningen
1962	Roll Damping Tests for a 103-ft T.S. Fast Patrol Boat for Malaya—Model 2324Z	Ned. Scheepsbouwkundig Proefstation—Wageningen
1962	Roll Damping Tests for a 162-ft W.L. Corvette for Ghana	Ned. Scheepsbouwkundig Proefstation—Wageningen
1962	Report on Experiments to Establish the Sensitivities of two Gyro Actuated Control Units as Used in Vosper Ltd. Roll Damping Systems	P. H. Main, B.Sc.
1962	Model Tests to Determine Various Aspects of Vosper Hulls 74, 77 and 78 in Calm Water and The Performance of Vosper hulls 77, 78 and 79 in Rough Water	Westland Aircraft Ltd. Saunders-Roe Division
1962	Model Tests to Determine the Performance of a 90-ft High Speed Planing Hull in Calm Water	Westland Aircraft Ltd. Saunders-Roe Division
1962	Model Tests to Determine Some Aspects of the Performance of Vosper Hull 83 in Calm Water	Westland Aircraft Ltd. Saunders-Roe Division
1962	Tests in Waves with 2 different high speed Corvette Models (Nos. 2324Z and 2439Z)	Ned. Scheepsbouwkundig Proefstation—Wageningen
1962	Fin Stabilisation tests for a frigate of the "Leander" type	Ned. Scheepsbouwkundig Proefstation—Wageningen
1962	Model Tests to Determine the Performance of Vosper Hull 71 (Ferocity) in Calm Water	Westland Aircraft Ltd. Saunders Roe Division
1963	Further Tests to determine the Calm Water Performance of Vosper Hull 71 (Ferocity) in the full scale trials condition	Westland Aircraft Ltd. Saunders Roe Division

Date	Title	Carried out by
1963	Methodical Cavitation Tests on Blade Sections Three Component Forces and Cavitation Pattern	Vosper Ltd.
1963	Resistance Tests on Vosper Model No. 89 (80-ft Fast Patrol Boat)	Ned. Scheepsbouwkundig Proefstation—Wageningen
1964	Comparative Resistance Tests on Vosper Hulls 74 and 83 in Calm Water. Brief Tests on Vosper Hull 74 in Head Seas Test on Vosper Hull 87 in Calm Water	Westland Aircraft Ltd. Saunders-Roe Divison
1964	Model Tests to Determine the Effect of Longitudinal Spray Deflectors on the Performance of Vosper Hull 83 in Calm Water	Westland Aircraft Ltd. Saunders-Roe Division
1965	Model Tests for Vosper Ltd. Calm Water Resistance Tests on Hull 91 and a comparison with Hull 87 in Head Seas	Westland Aircraft Ltd. Saunders-Roe Division
1965	Preliminary Report on Hull Bottom Pressures of HMTS *Trinity*	Vosper Ltd.
1965	Resistance and Stability Tests on a $\frac{1}{6}$ Scale Model of Vosper Model 90	Westland Aircraft Ltd. Saunders-Roe Division
1965	Further Tests on a $\frac{1}{6}$ Scale Model of Vosper Model 90	Westland Aircraft Ltd. Saunders-Roe Division
1965	Calm Water Resistance Tests on Hull 96	Westland Aircraft Ltd. Saunders-Roe Division
1966	Calm Water Resistance Tests on Vosper Hulls 83, 90, 93 and 99	Westland Aircraft Ltd. Saunders-Roe Division
1966	A Summarised Report of Structural Calculations on the "Flying Fish"	Vosper Ltd.
1964	Shear Stress in I-Girder With Side Strap on the Web	Vosper Ltd.
1965	Stresses in Modified Girders for the Wooden F.P.B. Design	Vosper Ltd.
1965	Hull Strength Weight Estimation of the 280-ft Corvette	Vosper Ltd.
1965	Structural Problems of Cross Stiffened Rectangular Plating	Vosper Ltd.
1965	Strength Calculations of Mark V Frigate	Vosper Ltd.
1967	Static Pressure Tests on Wooden Fast Patrol Boat Bottom Panels	Vosper Ltd.
1967	Model Tests of Roll Damping Fin and Bilge Keel Combinations for the "Leander" Frigate	Vosper Ltd.
1967	Model Tests on a 96-ft Fast Patrol Vessel	British Hovercraft Corporation (Saunders-Roe)
1968	Cavitation Tunnel Tests on Two Skegs designed for Hovercraft VT1	Vosper Ltd.
1968	Performance Investigation of 110-ft Patrol Craft for Peru	Vosper Ltd.
1968	Static Pressure Test on a Wooden Panel representing the proposed bottom construction for a new Fast Patrol Boat	Vosper Ltd.

PAPERS (LECTURES) READ BEFORE LEARNED SOCIETIES ETC. BY THE AUTHOR

Date	Title	Society
1953	High Speed Small Craft	Institute of Mechanical Engineers—Derby
1954	A Consideration of Some of the Principles Underlying the Performance of Planing Craft in Theory and Practice	Swedish Society of Naval Architects and Marine Engineers—Stockholm
1956	The Planing Performance, Pressures and Stresses in a High Speed Launch	Institution of Naval Architects—London
1957	Principles Governing the Planing and Control of Small High Speed Craft with Some Aircraft Analogies	Discussion Group of Sperry Gyroscope Co. Brentford
1958	Control of Roll Damping Fins (co-author R. H. Dadd)	1st Symposium on Ship Manoeuverability David Taylor Model Basin
1960	The Development and Running of the "Brave" Class Fast Patrol Boat	Swedish Society of Naval Architects—Gothenburg
1961	The Following Sea, Broaching and Surging (co-author G. J. Goodrich)	Royal Institution of Naval Architects at Gothenburg
1961	High Speed Planing Craft	Southern Joint Branch—R.I.N.A. & I.Mar.E.
1963	The Reduction of Roll in Ships	Southern Joint Branch—R.I.N.A. & I.Mar.E.
1968	Some Aspects of Planing Performance and Behaviour of Relatively Small High Speed Craft including Racing Boats and Fast Patrol Boats	Colloquium at National Physical Laboratory (Also submitted to 7th Symposium of Naval Hydro-dynamics—Rome 1968)

Lectures also given at:

Engineering Society—City & Guilds—London University
Eton School—Engineering Society
Manadon Naval Engineering College—Plymouth
St. Albans Engineering Society

Conclusion

As an engineer of sorts one must realise that one is growing older and that younger brains tend to take over where one is leaving off. However I do not find this disturbs me in the least. I find great pleasure in trying to keep pace with them or at least understanding what it is all about. There are even areas where the experience one has gained, sometimes the hard way, can be of help. I have to admit I like being with young people, they have such a refreshing enthusiasm and do not seem to mind hearing of one's many experiences in life. In fact they seem interested and I believe they are. If they are not I hope they won't show it, because to interest the young is a

Fig 66 The Author at his drawing board (circa 1966-7)

great pleasure to the old in years even if they do not feel old in mind and enthusiasm.

The title of this book does not allow me to discuss a number of things which have happened to me which might be interesting or even amusing. Nor does it really allow me to discuss my family, from all of whom I have derived the greatest pleasure in their company though they probably do not realise this. Growing old does not worry me greatly. I rather like the evening. It is a good time of day when the sharp ups and downs seem to be smoothed to an even and pleasurable tenor.

I owe much to the kindness of Sir David Brown our owner, under whom the firm is now prospering so startlingly well.

I have not mentioned 40 years of hunting in the shires where to sit a thoroughbred horse and ride across green fields and fences offers one of the most pleasurable sensations man can experience. This is not really much to do with engineering and it is probably an unusual pastime for engineers, but what a wonderful piece of engineering is a horse.

Index